THE MAIN PRINCIPLES

These are the three key ideas that we should focus on when thinking about cooking sustainably:

ZERO WASTE

ORGANIC AND SEASONAL

50% OF PRODUCE WITHIN 30 MILES

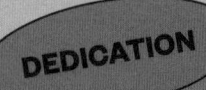

DEDICATION

This book is for my grandparents,
Grandad, Dad's Dad, Gargar and Jess,
who have passed on their baton
as we shall to our grandchildren.

NOTES

It goes without saying that all the
ingredients used in these recipes should
be organic, locally produced and
sustainably sourced as much as possible.
However, always use up what you have
before going out to buy something new.

When foraging for wild food, you must
be able to correctly identify what you are
picking, otherwise you should not eat it.

Both British (metric) and American
(imperial plus US cups) measurements
are included in these recipes; however,
it is important to work with one set of
measurements and not alternate between
the two within a recipe.

All eggs are medium (UK) or large (US).
Uncooked or partially cooked eggs should
not be served to the very old, frail, young
children, pregnant women or those with
compromised immune systems.

This book was produced ecologically using
a waterless printing process in compliance
with objectives of the Environmental
Protection Agency.

First published in the United Kingdom in 2021
by Pavilion
43 Great Ormond Street
London
WC1N 3HZ

Copyright © Pavilion Books Company Ltd 2021
Text copyright © Ollie Hunter 2021
Photography copyright © Louise Hagger 2021

ISBN 978-1-91164-178-0

A CIP catalogue record for this book
is available from the British Library.

10 9 8 7 6 5 4 3 2 1

Reproduction by Rival Colour Ltd, UK
Printed and bound by Leo Paper Products Ltd, China

www.pavilionbooks.com

Publisher: Helen Lewis
Editor: Cara Armstrong
Copyeditor: Alice Sambrook
Art Direction: Evi-O.Studio | Evi O
Designer: Evi-O.Studio | Nicole Ho & Laura Russeil
Illustrations: Evi-O.Studio | Susan Le & Kait Polkinghorne
Photographer: Louise Hagger
Photography assistant: Sam Reeves
Food stylist: Valerie Berry
Prop stylist: Rachel Vere
Production controller: Phil Brown

MIX
Paper from
responsible sources
FSC® C020056

Waterless Printing is
applied on this product

Join the GREENER REVOLUTION

30 Easy ways to live and eat sustainably

Ollie Hunter

PAVILION

CONTENTS

2 YOUR COMMUNITY

3 THE WORLD

WHAT IS THE GREENER REVOLUTION?

I accidentally became a chef really, not through ambition, but because of my love of nature. It is this passion that led my wife and I to buy a pub (a mad decision in hindsight), with the sole aim of making it the most sustainable pub in Britain (something we achieved in 2019). It gave us a place to be experimental in solving problems around food waste, leading to creative solutions and delicious, sustainable food. The greatest thing we have learned on our journey is that food is so intrinsically linked to nature and the natural cycles. We are all part of the constantly changing and evolving energy that is planet Earth. It is its own person, and we have been granted the fortune to be a part of its history.

My first book, *30 Easy Ways to Join the Food Revolution,* was dedicated to helping anyone become a sustainable food ambassador. I explained that everything we waste, we end up eating – for example, if we waste plastic it will enter the food chain by breaking down in the sea; fish then eat it and we consume the fish. If we spray pesticides on our vegetables, we end up eating those chemicals too. In this book, I'll be taking that message further and explaining how everything we waste in our daily lives, we also absorb in another way, whether it's the plastics from washing our clothes or the chemicals in the paint on the walls of our bedrooms. *The Greener Revolution* is a plan to help you become a sustainable ambassador in all areas of life.

Because food is an integral part of a sustainable life, and of my life, this remains a cookbook at its heart. There are new recipes, techniques and ways of thinking to help you eat more delicious, cheaper and more sustainable food.

Cooking is one simple thing we can all do, but I realize that suddenly transforming the rest of your life to become more sustainable isn't as easy as getting a new look at the hairdressers (although you could collect all your waste hair and use it in your compost as it's so high in nitrogen), so I've chopped up the solution into 30 bite-size 'ways', for you to digest slowly and make the changes as and when you can. These are separated further by being split into three chapters: Home, Community and World. The idea is that we start by making changes at home, then in our community and finally thinking about how our changes can affect the rest of the world. I'll be using the tree analogy (see pages 8–9) to show how these three stages of change are inextricably linked, and how they connect with one another in a very important way.

GETTING STARTED

We'll also be taking the three key principles of the Food Revolution (from my first book) and applying them here too. These are:

Zero waste Everything we do has the ability to produce zero waste. Either by finding uses for by-products or by recycling the waste properly back into the system that we all belong to. Zero waste is a way of making more with less.

Organic This isn't just a restaurant tagline or an extra 20% on your food bill, it's choosing to buy products whose raw materials haven't been sprayed with chemicals that kill our ecosystem from the bottom up. Organic produce can be affordable if we cook sustainably, and likewise choosing organic in other areas of life can be affordable if we live sustainably.

50% from within 30 miles If we all take responsibility for investing in our own 30-mile vicinity, then the whole world will work at its best and we will create a strong and positive future for our species. The motto to remember for this principle is 'connect globally, live locally'– everyone needs to do their job on a local level to ensure that we all flourish globally. Because everything we do impacts everyone and everything.

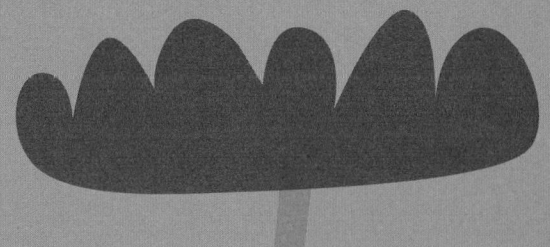

EVERYTHING IS CONNECTED, EVERYTHING IS ENERGY

I want to start this book with a tree. The same metaphorical tree I described in my last book that gave us the important motto 'connect globally, live locally'.

A tree shares its life with more than just itself – it shares it with the animals in its branches and the insects in its bark and leaves. It shares its life with the carbon dioxide it takes in and the oxygen it gives out, and with the water and the soil. It shares its life with bacteria and mould, with teeny-tiny organisms. It shares its life with the mycelium (a web of thin as thread fungal organisms). A tree is only able to flourish if each of these codependent relationships is working at its best. It needs them and they need it. These relationships range from the obvious happenings, to the microscopic ones and the ones that we haven't even discovered yet. This mutually beneficial transference of energy is like magic.

Humans are the same – we are intrinsically linked to everything on this planet. We are a constantly changing product of what's happening around us and what's happening in us. We rely on complex relationships with nature every second of every day. The energy that we give to the world will come back to affect us, as we are all part of the same ecosystem. Everything we do on Earth somehow impacts everyone and everything. If everyone in the world makes their homes sustainable on a local level, then each home is a strong and positive link in the global web of codependent relationships. If everyone makes their community sustainable, then the system becomes even stronger as a whole. If everyone made their vision of the world sustainable, then we allow ourselves to fully flourish as the wonderful and amazing species that we are.

The Greener Revolution is all about celebrating and making the most of this infinite connection that we have no choice over. We are all part of the same beautiful planet, and by making the energy that we transfer sustainable, we are making ourselves the best we can be.

WE ARE THE STORYTELLERS

If you feel concerned that a greener revolution might be a tall order for humanity, just remember that one of the most wonderful things about humans, which truly separates us from the rest of the natural world, is our ability to create stories. To allow our imaginations to wander away into other worlds and create visions and ideas that haven't existed before.

Everything that we've created on Earth so far started with a story – the plane is a story of us wanting to fly like a bird. A boat is a story of us dreaming that we could travel the seas like a great blue whale. A spaceship is a story we told ourselves about reaching the moon.

A nail is a story about building a home to protect the tribe. Electricity is a story about wanting to see in the dark. Families are a story about our genetics. Penicillin is a story about wanting the species to survive. Speech is a story about us wanting to communicate with each other. Clothes tell a story of our identity. Music is full of stories written by people wanting to express themselves, and when choosing what music to listen to, we apply these stories to ourselves. Art is another story about expressing emotion. The Internet is another story about communication. The Bible is a collection of stories about a God and his son Jesus. History is a series of stories about the stories we create. Contraception is a story about pro-choice. Borders and national identity are stories that never existed before humans created them. A government is a successful or unsuccessful story about representing the people within a story. Money is a story, full stop.

We have created so many stories, some truly wonderful, some not so good. Waste is a story about us not thinking of the future. Pollution is a story about damaging the natural world we live in. Climate change is a story about the possible end of our species.

We have created all these stories, but Earth will continue with or without them. So, we need to create new stories. We can create the solutions. We can create a better world. We can create harmony. We can live sustainably.

Because we are the storytellers and we have a new story to tell.

THE BIG QUESTION: WHAT IS ENOUGH?

There is always more, but what is enough?

I think this is probably one of the most important questions that we, as a species, need to start asking ourselves. What is enough food on my plate? What is enough TV to watch today? What is enough money? Do I have enough things? Do I have enough followers? Do I have enough friends? Am I enough?

I live in England, and you only have to do a small amount of travelling to realize how genuinely lucky people who live in England are. We have so many everyday luxuries compared to the rest of the world, which have all become normalized. To have any type of food at our disposal all the time is nowadays seen as a right, rather than a luxury. But do we need it? What is enough for our species to flourish?

A SMALL EPIPHANY

I had a small epiphany in the supermarket the other day. This moment of realization stemmed from something that made no sense to me at the time – when did we go from one sandwich being enough, to needing one and a half sandwiches?

You can now buy a pack of three sandwich halves in most grocery stores, which proposes two things:

That first sandwich was never filling, tasty or nutritious enough in the first place.

OR

That our diets and desires have at some point increased.

Or is it both?

If the food was tasty and nutritious enough in the first place, then we would only need one decent, beautiful soul-warming sandwich.

It suddenly became clear to me where we are going wrong: we are spending more time and money making more things that don't taste as good, or give us the nutrition we need. Hence, it will never be enough. It's maddening!

We will always be left wanting more if we just produce additional amounts of the same low-quality things.

So, what's the answer?

A MORAL RENAISSANCE

To find the answer, lets go back and look at how the need for one and a half sandwiches might have begun.

The cultural, political, scientific and intellectual advancements of the Renaissance Period gave us the belief that we humans are limitless in our capacity for development. This led to the notion that we should embrace all knowledge and develop our capacities as fully as possible.

After The Age of Enlightenment and the Industrial Revolution, the Western world found a determined will to succeed economically. This led to a linear fixation on a happiness derived from economic success – the idea that people with lots of money can afford whatever they want and are therefore happy. The thought process of modern day consumerism has therefore become: if I want something, I buy it because I can – even if I don't need it, understand where it has come from, know who has made it or what it is made of. More is more.

But life is not as linear as the Western world always sees it. To paraphrase some words from British Creativity expert Sir Kenneth Robinson...When we're teaching our children, let's not have history lessons and French lessons as separate classes – let's have a lesson about the French Revolution, whilst speaking French and, you know what, let's make a guillotine as well and throw some art into the equation – because the world is circular, not linear.

In other words, life is made up of many subjects, and it's far more beautifully complex than we have been used to allowing for. We must spark a moral renaissance by first acknowledging this in our schools, at work, in our homes and in our communities.

The Western world has given us so much, and we have achieved phenomenal amounts to evolve and progress the human race. But now I fear that linear progress is the very thing that is endangering us. It's the reason why consumerism is rife, why we are eating unsustainably and living unhealthy lives.

The problem is that we are using nature to evolve, rather than evolving with nature. We see food as a fuel rather than an integrated part of society. We are flooding to the cities and therefore see the countryside as just a place to grow our fuel.

So let's not do this anymore, because we can't afford to! I think it's time for the energies to rebalance – for a moral renaissance. Can we look to where people live simpler lives in the East for guidance on the things we may have lost – for spirituality, moderation and a different way of living? Can we enter into a truly wonderful time where East meets West in harmony?

Let's rekindle a positive relationship with nature and work to creatively solve all the problems presented, but this time morally and ethically. In simple terms, learn to enjoy life's luxuries and pleasures sustainably.

A few reasons why our current system can't continue:

☐ Over the last ten years we have produced more plastic than during the whole of the last century.

☐ People who live in high-density cities with high levels of pollution have a 20% higher risk of dying from lung cancer than people living in less polluted areas.

☐ People use cleaning chemicals at home that are ten times more toxic per acre than pesticides used by farmers.

☐ About 7 million premature deaths annually are linked to air pollution. That is one in eight deaths worldwide.

☐ Outdoor air pollution is responsible for the equivalent of up to 36,000 premature deaths a year in the UK.

☐ Two-thirds of all UK car journeys are less than 5 miles (8 km).

☐ 1.8 billion disposable nappies are thrown away in America every year.

☐ Air pollution costs the UK £20 billion a year.

☐ Earth now has 46% fewer trees than it did 12,000 years ago.

☐ Only 9% of all the plastic ever made has been recycled.

IT'S NOT YOUR FAULT

Despite these slightly scary facts, if you've read my first book, you'll know that I started with the same reassurance there too. We know that there are problems in the world, and sustainability is one of the largest. The environment is changing faster than we can predict and the stark truth of it is pretty terrifying.

BUT, it's not fair that you should have all that pressure on your shoulders. Why should we have to individually burden ourselves with everything that is wrong in the world? We shouldn't. Life is difficult enough as it is. The important thing is to understand why we have to make changes.

You can make a huge difference in the world. We all can. And we can have fun doing it. We are all entitled to enjoy life, to have happy, fulfilling relationships, a decent quality of life, support for our health, good education, shelter and nourishing food and drink. As long as our quest for these things is sustainable and does not upset the balance of Earth and our survival here, then life can be wonderful and beautiful.

So... no more negativity about the end of the world, let's live, cook and eat sustainably.

Join the Greener Revolution.

#30Green

There's hope!

CHAPTER 1

Is there anywhere as important as your own home? How wonderful, how blissful, how comforting to have your own sanctuary from the outside realm. It's your very own world that you've created to express yourself.
It's a house for your story.

YOUR

HOME

If the world is our metaphorical tree (see pages 8–9), then it could be said that each home is like an individual mycelium. If we all make sure that each every mycelium within the larger web is operating sustainably, then the whole system could work towards the same purpose more easily. Lots of small changes can make a big change.

THE GARDEN

Grow your own oxygen (and food)

The city of Amsterdam is famous for all sorts, but what I hadn't anticipated when I visited is just how much beautiful greenery there is everywhere. Every household seemed to have plants and trees growing in front of its doors. Streets were littered with pots, barrels and even unused baths, all containing plants producing oxygen for the city and food for people's homes. The residents of Amsterdam don't let being in a built-up city stop them from planting something in every space available – we should all be taking inspiration from them.

Plants add colour and positivity to every day, they provide nectar for pollinators and create oxygen for you and me. Being passionate about food, I've always found growing edible things particularly rewarding. Herbs are one of the best seeds to sow as they will sit patiently on your windowsill or doorstep, just waiting for you to use them time and time again. Many herbs also create flowers, which are fantastic for bees and other pollinators and also for us to eat – try vibrant pink chive flowers sprinkled over eggs, rich purple sage flowers in a ricotta ravioli, mustardy nasturtiums in a potato salad, violet lavender-infused sugar with an apple tart and yellow dill flowers sprinkled over asparagus and hollandaise. Simply get yourself a container, fill it with soil, sprinkle with seeds, water regularly as per the instructions and enjoy the freshness and flavours.

Most vegetables are easy enough to grow outdoors or indoors in containers. Good options to try if you are new to growing veg include salad leaves (normally one of the worst recipients of pesticides), beans (French, broad/fava), cannellini, borlotti), peas (mangetout, petit pois), courgettes/zucchini (you can eat the flowers, fruit and stems), radishes (leaves as well) and tomatoes.

Plant-growing tips

Unusual spots Opportunities for growing space exist everywhere once you start looking. Windowsills, window boxes, pots in your porch, on your steps, on your railings, hanging baskets, roofs, in lattice woodwork, frame shelves, a table centrepiece, hallway feature, bathroom greenery... the list goes on.

Think vertically Space is limited in our constantly expanding urban jungles, so look to the skies for inspiration and grow vertically. Walls are a very underused space and there are many plants like fruit trees, fruit bushes, beans, tomatoes and peas that don't need a lot of ground space but can maximize the space you have in the air.

Upcycle The trend of upcycling has lots of benefits and you don't need to be a hipster to enjoy it. How about using old pallets chopped in half and stacked on top of each other to grow herbs poking out through the gaps? Old disused wooden ladders can be loaded with plants on every step. Wooden fruit crates lined with wood can be filled with soil. Try wheelbarrows, old food tins, old teapots or teacups – anything can be a home for a plant!

House plants

House plants reduce carbon dioxide and increase oxygen levels within the home, bringing life and energy to any room. They filter dangerous pollutants like nitrogen dioxide from the air using their leaves, are said to balance the humidity, reduce sound levels, increase positivity, reduce stress and anxiety and improve concentration among other things. It's hard to think of a reason not to get one! Imagine, just by getting yourself a leafy friend who only needs watering every now and again, you could be sleeping in a room with more oxygen.

Easy-to-care-for house plants include spider plants, parlour palms, rubber plants, Boston ferns, umbrella plants, corn plants and broadleaf lady palms (just remember to check to make sure the plant you choose is not poisonous to your pets). Next time it's your birthday, ask for a plant and save on the wrapping paper as well. Make looking after it a part of your daily routine, like brushing your teeth, feeding a sourdough starter or making a cuppa. And buy one type of plant first to see if it suits your room before turning it into a jungle!

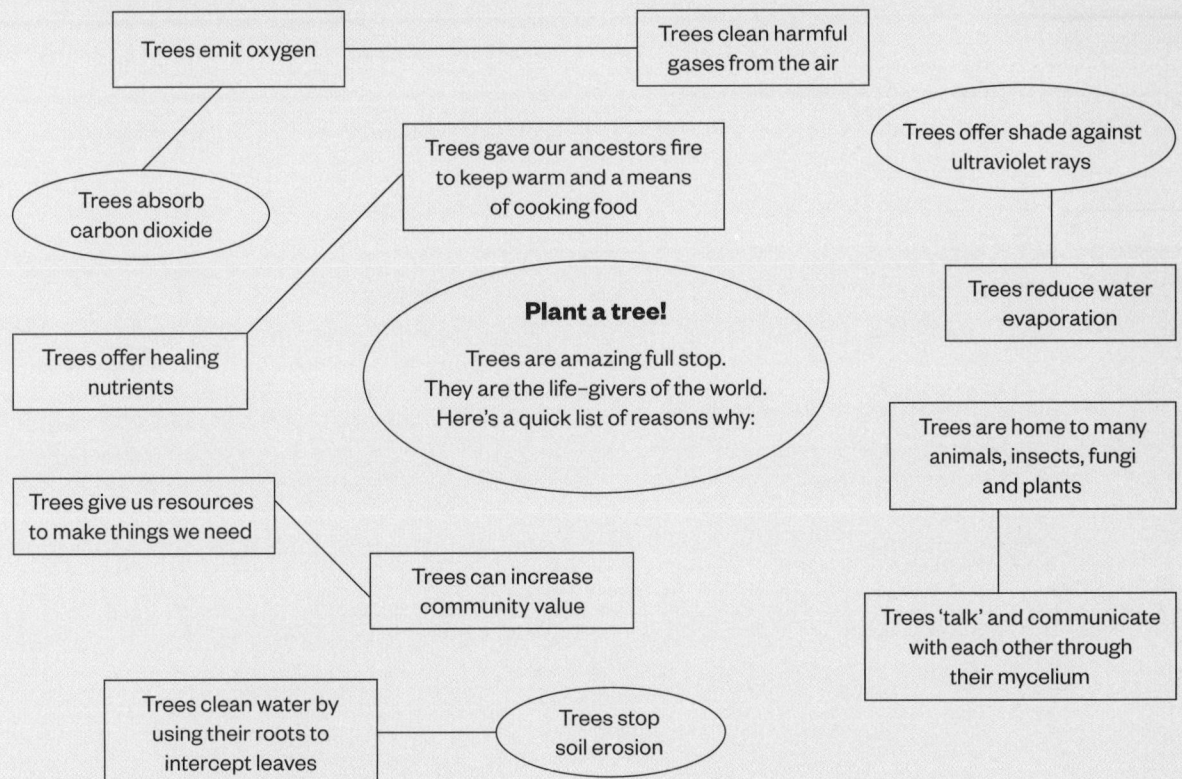

Every tree we plant helps to absorb the dangerous amount of carbon dioxide in the air, and is one small step closer to putting the brakes on climate change.

If you're lucky enough to have the space to plant a fruit or nut tree outdoors, then you will also reap the produce rewards for a very long time. If you don't have access to an outside space, then grow a miniature tree inside and let it reduce CO_2 and air pollutants indoors. A good place to start your indoor arboretum adventure is with a weeping fig tree, a fiddle-leaf fig tree or a citrus tree if you have a particularly sunny spot.

The idea is to try and plant to replenish what we consume. If there is less dependency on dwindling resources, it should mean less violence between people fighting over what is left. Plant more and live more peaceful lives.

THE KITCHEN

2

The kitchen is one place in particular that single-use plastics shouldn't be anywhere near, because of the transfer of plastics into our food systems. It didn't used to be this way – in years gone by, most kitchen items were made out of durable natural products like wood, ceramic, glass and metal. I can still remember the smell of my grandmother's kitchen, the feeling of her ancient but sturdy wooden spoons and the character of those beautiful cast-iron pots that seemed to have cooked a thousand meals for a thousand people. But somehow, over the past 50 years or so, plastic has managed to creep its way into many areas of the kitchen like a slithery snake (or perhaps a slippery fish). It's probably got something to do with the throw-away culture that we've become used to.

You might argue that a really good plastic spatula can do things a wooden spoon can't, such as get every last bit of sauce out of the saucepan (which is better for zero waste). In cases such as this, our aim should not be to buy more, but to reuse and recycle the plastic items we already have. We can't unmake them, but we can put them to good use.

Going forward into our greener revolution, we need to heed the phrase 'measure twice and cut once' or to put it another way, 'think twice, buy once'. For example, buy one decent knife, look after it, and it'll last you a lifetime. Really think about any purchases you make for the kitchen and imagine living with that product forever. Investment pieces might seem more costly in the short term, but investing in quality equipment will save you money, time and effort in the long run. Instead of buying new, you can also use second-hand pieces of equipment handed down from family or friends or found in charity shops or online. I still use my great uncle's ice-cream machine from the 80s, and it works perfectly. Another option is to ask neighbours, friends and family to borrow equipment that you don't need to use all of the time.

The lists on the right detail everything you should need for a fully functioning eco-conscious kitchen. You may not need everything listed, and be sure to get all the use out of the items you already own before buying anything new.

General equipment

- Wooden spoons and chopping boards
- Cast-iron cookware
- High quality metal knife (one good one is better than a set of average ones)
- Glass measuring jugs and dishes
- Mechanical kitchen scales
- Rubber mats (instead of baking parchment)
- Durable organic cotton tea towels (kitchen cloths)
- High-quality oven gloves that will last
- Reusable tablecloths or napkins made from organic cotton or hessian (instead of paper ones)
- Used enamel items

Electrical equipment

- Fridges – smaller fridges can save money and electricity. People often have a lot of empty space in their fridges and keep lots of food in there that doesn't need to be chilled – some fruits for example
- Freezers – whatever you do, don't downsize on these! Freezers are one of the greatest tools in a sustainable kitchen. They are useful for batch cooking, storing whole animals or keeping gluts of produce. Reaching for the homemade burger in your freezer is far better than a ready meal or a rip-off takeaway.
- Pressure cooker – an energy efficient way of cooking
- Induction hobs – again these use less energy than gas or electric (note that with these you can only use stainless steel, cast-iron or pans with an induction plate built into the base). Alternatively, you could switch to a green gas company for your gas hob
- Dishwashers – if you need a dishwasher, ensure you have a full load before using it. Skip the pre-rinse stage and turn the heat down. Try to buy dishwasher powder sold in a cardboard box
- A domestic flour mill (optional but very useful)

Storage

- Cotton bags
- Paper bags
- Glass jars for dried goods with bamboo or cork lids
- Glass bottles for keeping water in the fridge
- Kilner (mason) jars
- Storage containers (rather than clingfilm/plastic wrap)

Takeaway items

- Keep cups
- Metal water bottles
- Metal or bamboo lunchboxes

Cleaning items

- Wooden or bamboo dish brushes
- Reusable cloths (please never use disposable wipes! They contain huge amounts of plastic and are not biodegradable)
- Make your own cleaning spray (see page 23)
- Metal and wooden dustpan and brush
- Kitchen buckets in sinks (so you'll use less water)

The Kitchen

How to shop for food

Buying in bulk is the way forward – it means there is less packaging, less energy spent on transporting the goods and it's cheaper in the long run. Visit butchers, fishmongers or local farms to get the best deals on bulk buys. Another great thing to do is visit zero waste shops and take your own containers to refill stations. Think about each item on your shopping list and consider what might be the most sustainable way of buying it, for example, getting your milk delivered the old-fashioned way in reusable glass bottles or switching to loose-leaf tea and loose coffee rather than pods. By changing the way we shop we can create demand and change the systems already in place – a great example of this change in action is the fact that supermarkets are slowly waking up to calls for less packaging as a result of demand from shoppers.

How to keep leftovers

There is a growing demand to replace clingfilm (plastic wrap) with beeswax wraps. Although the latter product is far more natural, I hesitate to endorse the continual demand on the struggling bee industry (for more on this see page 73 of my first book – *30 Easy Ways to Join the Food Revolution)*. Right now, our bees need to be left alone as much as possible to strengthen their own ecosystems. There are plenty of other alternatives to clingfilm, like covering a bowl with a plate, using a storage box or wrapping something in a damp cloth or putting it in a paper bag.

I also think we could put more effort into making reusable wraps and foils out of plant resin – and by demanding more plant resins we demand more plants, which means more oxygen. If you manage to find some plant resin wraps to buy, just make sure they are soy-free, palm oil-free and locally sourced before purchasing.

Pets and their food

In the United States alone, roughly 20% of all meat consumed is by pets. There are roughly 9 million dogs in the UK, which is almost the entire population of London. So please don't buy plastic pet-food pouches. Don't throw away your own leftover food, but feed what is suitable to your pets instead, or ask trusted friends and neighbours for their leftovers. And when you're buying the whole animal from a butcher, ask for the innards and pets will love this treat. Look to supplement with eco-conscious brands of pet food and those that use sustainable packaging and careful sourcing. We are all part of the same ecosystem, so we should absolutely be aiming to use the same principles for pet food as we do for human food – zero waste, organic, and 50% from within 30 miles.

Natural cleaning products

My grandmother always says 'Champagne should be on the NHS' (as it's medicinal). Well, although I agree with her, I also think cider vinegar should be dosed out liberally because it's wonderful for so many things. In England, we have so many apple trees that give back year after year. In the same way that European countries have cooperatives for olives, wine and other local produce, I think England should have localized apple cooperatives where we can bring our own local apples to be turned into a variety of products, from vinegar, to juice, to molasses to compost and cleaning products. If we did have localized pressing stations, then we may never need to buy cleaning products full of chemicals again!

When chemical-laden cleaning products go down the drain, they enter the water supply. This not only affects the environment, but over time they can build up and become toxic to us and our bodies too. Natural cleaning products work well as disinfectants, deodorizers and grease cleaners because the acid in the vinegar chemically changes the proteins and fats that makes up viruses and bacteria, destroying their cell structures. They're not quite as effective as very harsh chemicals, but we're not trying to wipe out all life on Earth, just clean. If you really want to get something super clean, then the best way to do it is using hot soapy water and good old elbow grease. Scrub the area hard, then spray/soak with pure vinegar, leaving that on for 10 minutes before wiping it away.

If you don't have access to cider vinegar then any type of vinegar can be used in all the cleaning recipes to the right. Don't forget to also regularly clean your cloths or sponges themselves – obvious maybe, but so important.

Sprays for the bathroom and/or kitchen
Simply mix 1 part vinegar with 1 part water. You can leave naturally-scented ingredients like rosemary or lavender in the vinegar to infuse for 10 days before using, or add essential oils in a natural fresh scent of your choice, such as apple and mint. Leave for 1 week, then strain (if needed) and pour into a reusable spray bottle. You can also replace the vinegar with citric acid powder if you like, but use warm water to dissolve the powder.

For mirrors   Make a cup of tea! The astringency of a well-brewed cup of black tea is the perfect thing to cut through grease and dust. It's also great on other shiny surfaces like windows or wood floors.

For tough surface stains   Make a thick paste with equal amounts of bicarbonate of soda (baking soda) and water. Scrub this into the affected area, leave it overnight and then wipe away the next day.

For carpet stains   Generously sprinkle bicarbonate of soda (baking soda) over the stain, spray with some hot water and leave for a minimum of 4 hours. All you need to do is vacuum it up later or the next day. Alternatively, you can mix together 1 part washing soda (sodium carbonate) with 1 part vinegar and 10 parts water and clean as normal.

For wood   Mix 1 part vinegar with 30 parts water and rub on with a soft cloth. Or use 1 part vinegar, 4 parts oil and 30 parts water. Or, for really tough stains, mix 1 part vinegar and 1 part bicarbonate of soda (baking soda) to form a paste and then scrub (you might need to re-oil the wood after this one).

For drains   Add 250 g/9 oz of bicarbonate of soda (baking soda) to the drain, leave for 1 hour, then add 250 ml/8½ fl oz/1 cup plus 1 tablespoon of vinegar. Leave to sit for 12 hours, then rinse thoroughly with water.

Tough pans   Sprinkle over 2 tablespoons of bicarbonate of soda (baking soda), then add 2.5 cm/1 inch of water and simmer gently for 2 minutes on the hob (stovetop) before wiping away the dirt.

THE BATHROOM

The bathroom is one of the main culprits of the endless demand for products contained within single-use plastics. I'm afraid a lot of the responsibility for this, as in many other cases, is down to the advertising and marketing campaigns that make us feel as though we *need* these products. In actual fact, most toiletries are full of things we don't need, such as preservatives (so products have long lives to sit on shelves), fragrances (to hide all the chemicals) and surfactants (these give instant foaming effects, but actually damage aquatic ecosystems). If we are what we eat, then it follows that we are what we clean ourselves with too.

Our bodies have not spent thousands of years in evolution to create systems that are bad for us – they are great at self-regulating to keep our skin and other areas in the best possible condition all on their own. How many of the available scrubs, creams, lotions and potions can we really not do without? Many of them are probably just confusing the natural processes in our bodies. Try stripping your list of toiletries right back and see what you can do without. Plus, it'll be so much cheaper that you could probably afford an extra holiday within the year.

Refill stations

Without a doubt, the first thing you can do to make your bathroom a greener place is start using refill stations. Shopping for bathroom refill products is actually easier than doing it for food, because we probably already have suitable empty containers from previous products and we don't need to find a huge variety, so there's no need to bring loads of containers. Key items you can refill on include:

- Shampoo and conditioner
- Body wash
- Moisturiser
- Soaps
- Shaving creams
- Toothpaste (in both paste and pill form)
- Cleaning products

Other eco-friendly bathroom switches*

- Wooden or bamboo toothbrushes
- Reusable cotton or bamboo make-up remover pads
- Recycled loo paper – there are some fantastic brands out there that sell recycled toilet paper packaged in recycled paper, who don't use chemicals and who use their profits in charitable ways, like building loos for people who need them
- Shampoo and bodywash bars – the production of liquid soaps requires five times more energy than soap bars, and can use up to twenty times more packaging
- Palm oil-free soaps
- Plant-based products
- Bath towels and bathmats made from organic and sustainable materials
- Reusable washcloths – they wash really easily and colour coding them is a great way to know whose is whose

* As with the kitchen items, don't buy anything new until you run out of old products, and buy in bulk if possible to help reduce packaging and transport.

Things to think about installing

Some of these things are cheap and easy, others are expensive and require work. Don't feel like you need to install all of these, but pick what's best for you, save up and/or wait until something needs replacing before you make a switch. Build up to these moments and make sure you congratulate yourself when they happen – investing in living more sustainably is such an achievement.

- Put two sand-filled containers in the toilet cistern – this reduces the amount of water that the cistern needs to fill up, so each flush uses less water. Experiment with how large the containers are to control the amount of water you need
- Make shelves out of reclaimed wood
- Use reclaimed glass toothbrush holders
- Use locally made furnishings, or even better products made within 30 miles of where you live
- Use natural products like stone, cork, and 100% natural rubber
- When you need a new bathtub, get a metal one – acrylic or fibreglass normally end up in landfill
- Motion sensor taps
- Low flow taps and showers
- Low energy extractor fans
- Shower aerators – these can reduce the amount of water to half without loss of pressure
- Waterless loos
- Compost loos
- LED lights
- Underfloor heating – this avoids overheating and wasting energy

Cleaning thyself

What is enough? Do I really need to wash myself all over with soap every day? I actually started to reduce my soap usage a few years back. Now, I wash my body once or twice a week with soap. Clean, hot water is actually very good at cleansing the body, because most of the time we don't need to be super clean. Obviously, I do the important bits daily (especially hands, which are used for everything). But washing all over with lots of soap too often can actually disrupt the natural PH levels in our skin.

Our top dead layer of skin cells contains lipids (a sort of fat) that protects the healthy new skin underneath. Constant cleaning with soap breaks down the protective cells on top, which puts the body into a continuous state of repair through excessive oil production. The likelihood is that it won't be able to keep up though, and skin will become dry. The resulting irony is that through overuse of soap, we then feel the need to repair ourselves by using lots of another product to replenish moisture. Advertising agencies and skincare companies stand to profit from our expensive personal hygiene habits, but it is a pretty inefficient system for us and for the environment.
The same goes for shampooing our hair. Excessive use of shampoo every day will dry your hair out and make you reach for lots of conditioner. Instead, use less good quality products with natural ingredients and let the natural oils on your scalp do the job they are meant to do.

Everyone is different, so I cannot prescribe a daily routine, but just questioning your own personal habits and keeping an eye on the quality and amount of products you use will help you and help the world.

Homemade shampoo

Making your own shampoo is, of course, the most Earth-friendly way of washing your hair. There are many books and online tutorials out there dedicated to offering different homemade shampoo recipes suitable for different types of hair, so do your research and find one that works well for you. It may take a little time for your hair to adjust to the new routine, so I recommend alternating homemade shampoo with your regular shampoo every few uses and building it up from there.

Plastic-free periods

There is roughly 200,000 tonnes of menstrual product waste a year in the UK, which equates to about 200 kg of waste per person in their lifetime. I'm told that periods are difficult enough without having to contend with the guilt of all that single-use plastic too! Thankfully, there are many companies out there producing innovative reusable and/or plastic-free options that aim to be more comfortable for you and better for the planet. Whatever you decide to go for is rightly a very personal choice.

Menstrual cups   Made of medical-grade silicone, the idea is that these cups can hold more blood than other methods and they are a cleanable, reusable option. They require a small investment, but this pays off very quickly when you think you could save yourself the purchase of approximately 11,000 or more disposable sanitary products in a lifetime.

Reusable period underwear   Avoid those microplastics that exist in tampons and pads and opt for a pair of period pants instead. Depending on the brand, you can wear them for up to 12 hours and they can hold up to four tampons' worth of flow. They're leak-proof and are machine washable. Invest properly in a good pair, firstly because you deserve it and secondly because it will save you loads of money long term.

Reusable pads   Did you know that 90% of a normal menstrual pad is made of plastic? Sustainable versions of these are made from organic cotton, which is what should be next to your intimate areas rather than plastic and chemicals. This can save you itchiness, soreness and discomfort. Invest in quality period products for yourself and for the planet.

Charities   There are some wonderful charities out there which aim to bring safe sanitary products to everyone and train people to make their own reusable sanitary products. If you see these charities around, please think about donating.

#letstalkaboutperiods

#periodswithoutplastic

#plasticfreeperiods

#periodsareforlifenotjustforchristmas

#periodpresent

THE BEDROOM

4

We spend, or should spend, almost a third of our lives sleeping, relaxing and breathing in our own bedroom. It's like your own personal microclimate – everything in the bedroom is a part of you and you are part of it. So, if you love yourself then you should love what's in your bedroom too. Wool, organic cotton, bamboo, jute, seagrass, sisal and coir are all great materials to have around you. Again, try to use what you have first and don't buy anything new unless necessary.

Things to use/have

- Duvets should be made from natural, organic materials to avoid toxins and polymers. If you can, find one that's made locally and/or using sustainable energy
- Wool blankets instead of polyester
- Good-quality bed sheets – organic cotton and bamboo are fantastic options that will last
- Organic cotton or bamboo pyjamas (or let your skin breathe without pyjamas)
- House plants (see page 19)
- Non-toxic paint (see page 36)
- Second-hand furniture
- Home-dried herbs – lavender, chamomile and mint are wonderful to calm your senses

Things to think about installing

- Eco mattresses – made with organic, natural, toxin-free materials. Look for eco-friendly companies that have provenance, are charitable and you can buy directly from
- Beds made from reclaimed materials – second-hand is always a great option
- Upcycled furnishings – I've used old apple crates for my bedside table, it holds my books nicely
- Wood floors and rugs made from natural materials rather than underlay and carpets (carpets and underlay will inevitably go to landfill one day, where they will sit for a long time decomposing and releasing volatile organic compounds – not the good kind of organic)
- Curtains made from natural materials like linen, canvas, organic cotton or hemp and coloured using natural dyes
- LED lights

Keeping your bedroom fresh

- [ ] The simplest technique is just to open those windows and let the air flow. Good air flow is good dream flow

- [ ] For unpleasant smells, sprinkle some bicarbonate of soda (baking soda) on surfaces like mattresses and carpets, leave for as long as possible and then vacuum

- [ ] Use old unwearable clothes to clean curtains or blinds with a solution of 1 part vinegar to 1 part water

- [ ] Don't forget to flip your mattresses every 6 months

- [ ] For washing sheets and laundry, see page 29

Condoms

You know that moment when you're a teenager and your dad or mum starts talking to you about the birds and the bees and it's just awkward.

This is how these conversations should go from now on: firstly, USE PROTECTION. Secondly, WEAR SUSTAINABLE CONDOMS.

Obviously, the most important thing is that the condoms work in the first place, so buy from reputable companies. Then you can start looking at whether they use Fairtrade rubber, are non-GMO, organic, vegan, nitrosamine-free, paraben-free, free from fragrances and are charitable in their profit donations.

The same goes for lubricants and all that. On our moral renaissance theme, I am sure more and more sustainable solutions will appear over time, even from the moment I'm writing this to the publication date, because we are advancing very quickly. Keep an eye out for what's new. Never stop asking questions to find solutions.

How to sleep

Within the context of living sustainably, I think it is worth talking about sleep and how wonderful and important it is for us. Our bodies are ecosystems in themselves. Amazing products of evolution, when it gets dark they produce the hormone melatonin so that we begin to feel sleepy. This natural cycle happens every day.

My wife and I once stayed with a couple in Calabria, Italy, on their farm. They were completely off-grid – no electricity, no hot water unless we boiled it and no heating. They lived from the land and it was beautiful to be a part of. But because they had no electricity, they went to bed as soon as it became dark. It was quite funny to watch, as soon as the sun disappeared, they would become walking zombies. They had returned to their natural sleep cycle and it was powerful. Big up the melatonin. We followed their example and slept well.

I'm not suggesting that we should all 'lights out' as soon as night arrives, but your circadian rhythm works best when you have healthy sleep habits. Get into the routine of slowing down and having some calm time to reflect before bed. Avoid bright screens at least an hour before you turn in; dim the lights and let go of brain stimulation. Slow down the breathing. Turn off all the lights when you sleep.

And then in the day, walk to work in the morning, get outside on your breaks and let as much natural light into rooms and offices as you can. Live the day in daylight.

The more sustainable your bedroom, the less chemicals and interference you'll have during your sleep. Live sustainably to sleep better.

> **Tip** There is lots of evidence that we should be sleeping in cooler rooms due to the fact that we evolved for a long time sleeping without central heating. Try sleeping in a room between 15–18°C (60–64°F). Not only does the cooler temperature signal to the brain that it's time to sleep, but it also increases the body's energy expenditure which is good for our metabolism. This will save on energy and bills and probably help after eating a large meal.

THE WARDROBE

5

If we're going to change the way we look at everything and take on all these different industries in order to start a greener revolution, then the fashion industry is an important one that we need to talk about. Clothes are part of our identity; they represent what we want to say to the world and are a statement of what we believe in. But if we believe in the survival of the human race, then we need to change the fashion industry completely, and somehow do this without removing its soul – without removing the authentic creativity and passion that exists within it.

Many of our clothes these days contain plastics like polyester, nylon, acrylic and polyamide. In the same way that the food industry learned to bulk out foods with cheap fats, carbs and sugars, the fashion industry did the same to our clothes with cheap plastics. All these things are connected, to the point where what you wear affects what you eat and how you breathe every day. The microplastics in our clothes are so small that they can enter the water system through our waste treatment centres, then they get eaten by fish in the sea. They also enter the air and float around, for us and animals to inhale. Simply by washing clothes that contain plastics, we are polluting the oceans and the fish that swim in them, which we then eat. Simply by tumble drying clothes that contain plastics, we pollute the air that we and animals breathe in.

But do not panic! Certainly don't throw away all your old clothes and spend money on new clothes. We don't want to take away from the fact that fashion shapes culture, but instead we want to change the way we think about fashion. In the same way that thinking about food in a sustainable way leads to more delicious, nutritious and cheaper foods, thinking about the fashion industry sustainably will lead to higher quality clothes that last longer, are more creative and won't pollute the air or oceans.

First up, tackle this problem from the top down by signing petitions to put pressure on large companies to change how they manufacture clothes, what materials they use and what responsibility they have for waste. Secondly, make the following simple everyday changes to how you wash, wear, recycle and buy new clothes. It's amazing what little changes across the whole world will do, and the fashion industry is no different.

Washing clothes

Up to 25% of the fashion industry's carbon footprint comes from washing clothes. In order to combat this:

- Try to wash clothes less frequently
- When you do wash, ensure there is a full load – clothes are less likely to shed their microplastics and less energy is used on multiple washes
- Wash at low temperatures and on slower spins (there is often an 'eco' setting)
- Air dry your clothes rather than tumble dry as much as possible
- Buy special 'balls' to put in your washing machine that trap microfibres including the plastic
- Don't use fabric softeners! These are bad for clothes and very bad for the environment. Modern detergents are pretty good at what they do, and they don't leave clothes rough and scratchy. Fabric softeners embed themselves into the clothes, leaving us open to a myriad of chemicals and a far more flammable wardrobe. Many contain palm oil and even some animal fats
- Try adding 250 ml/8½ fl oz/1 cup plus 1 tablespoon of vinegar to the rinse cycle, this naturally breaks down stains and chemicals

Wearing clothes

Wear the clothes you like until they're completely unwearable, then use them as a cloth. I still have a jumper from when I was 14 – it was my favourite back then and I think it has grown with me!

Recycling clothes

You can upgrade broken, tired or old garments and give them a whole new lease of life just by changing a few simple details. Adding buttons, stitching in a pocket, dyeing a new colour (using non-toxic dye), making dresses from large T-shirts, gloves from sweaters... the possibilities are endless. Show off the clothes that you have made yourself and share sewing skills with friends.

And if you don't want to wear them any more, turn clothes into bags, teddy bears, cushions, blankets, a dog bed cover... the list goes on.

It's not what you consume that defines you, but what you create.

Buying clothes

Buy quality clothes that will last longer

Buy Fairtrade clothes

Buy organic clothes

Buy recycled clothes

Buy re-spun clothes

Buy clothes made from wool (it helps to fertilize the soils)

Buy clothes made from linen (the linen or flax plant needs very little to thrive and improves the soil)

Buy clothes made from new technologies that use wood cellulose as this also promotes more tree planting

Buy clothes that are made from recycled materials, even if nylon and polyester

Buy from charity shops

Buy clothes from companies that promise you free repairs

Buy clothes that are made by artisans to promote employment in rural environments

Buy clothes from companies that use water responsibly

Buy clothes from companies that use renewable energies

Buy clothes from companies that are metal conscious

Buy clothes from a circular economy business (see page 73)

Buy shoes to last for life. Vegetarian leather is a very good option

Let's not make it a class issue

It's important that we don't make these new sustainable practices a problem of class. I believe in purchasing well made, long-lasting, organic clothes, but it's not possible to afford these all the time. Buy fewer items per year and spend more on each one. Or visit charity shops, recycle businesses and second-hand stores. Spend money on things that can change other people's lives.

IN-HOUSE SYSTEMS

Let's go back to our tree concept (see page 9): connect globally, live locally. Connecting globally is about sharing the same goals and vision of the future, so when we install in-house systems to make our homes more sustainable, we trust that we are doing it for the whole species, as well as for ourselves. Also, the more we do on a local level, the more power we have individually and within our communities. I ended my last book on food sustainability by saying that cooking your own food gives you control because you grow it, you cook it, you eat it – and you waste less. Well, what systems can we install in our own homes on a 'local' level to regain the control we have lost to larger corporations?

At the pub, we introduced an in-house system to create or own sparkling drinks, which reduced the number of bottles we were buying, saved energy on transport, reduced waste and saved money. It also made the team happier because they had to carry fewer bottles up and down the cellar steps. One change = four happy results. The general rule is the fewer processes involved in the system, the less waste we create.

What other in-house changes can we implement to reduce energy and save money?

Compost heaters

First invented by French farmer Jean Pain back in the 70s, this wonderfully natural idea is a perfect way to provide heat in homes, hot tubs, greenhouses or sheds during the winter. A compost heater is made by placing a coil of water pipes in a large structure containing woodchips, sawdust and other compostable materials such as grass clippings and leaves. The long composting process of the wood chips and other items creates heat, which warms the water in the pipes. Once the materials have fully composted, you can then use them on your garden.

Heat exchangers

This efficient system is designed to take excess heat from warm rooms and transfer it to other cooler rooms that need it. If your home is designed well enough, it means that you could do without central heating or at least save lots of energy and money on your heating bills.

Home batteries

There is still a huge amount of technology to be discovered when it comes to storing energy, but home batteries are a great investment and give you more independence. As well as reducing the amount of money you give away to energy suppliers, they also relieve stress on the national grid, so you're helping your country as well. You can charge them in any way you like – by generating electricity from solar power, wine or even buying electricity at cheaper times throughout the day to use as and when you need.

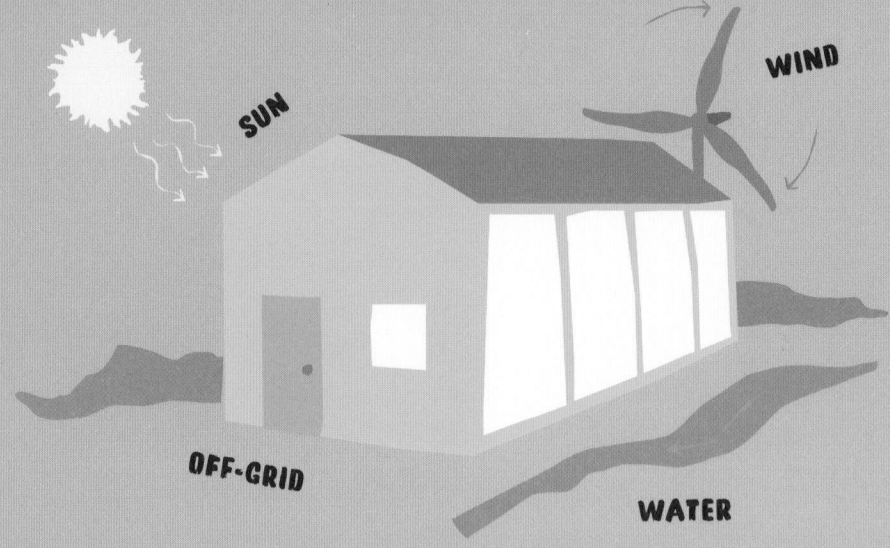

SUN

WIND

OFF-GRID

WATER

E-water

This is a clever new technology that works by passing a salt solution through an electrolyzer that separates the ions within it, producing two oppositely charged solutions. The positively charged solution acts as a sanitizer and kills viruses and bacteria. The negatively charged solution is a pure liquid soap, perfect as a general cleaning liquid. The two solutions revert back to their natural states after a period of contact with organic matter. It's a brilliant bit of natural chemistry and a sustainable alternative to products made with artificial chemicals. At the moment e-water is mostly used in commercial premises, but I think it's important to mention it so that we can create a demand for a version of this in our homes.

Other energy systems we can install at home

- ☐ Smart thermostats
- ☐ Heat pumps
- ☐ Solar panels
- ☐ Smart meters
- ☐ Electric heaters
- ☐ Log burners with back boilers

Earthships

These ultimate off-grid eco-homes require land, so of course are not possible for everyone. However, they are inspirational visions of how we can optimize our homes and communities to make them more efficient. The walls of Earthships are built with natural and recycled materials, all the energy they require is harvested from the sun or wind, rainwater is harvested and purified for all water-based needs and they have their own self-contained waste and sewage facilities. There is usually a south-facing glass façade housing a purpose-made garden for an in-home organic food supply. In our moral renaissance (see page 12), we should start to really think big about how we can improve building design – and the Earthship movement is a great example of how to use nature to the best of its ability. There is continual research as to whether all aspects of these designs work as well as they could or if they could be improved on, but the vision is what we should be aiming for.

The more we create ourselves, the freer we will be from social, political and economical controls.

DANGER –
HIGH VOLTAGE!

On Monday mornings at the pub, in between prepping food, calling recycling companies who haven't turned up and answering emails about whether we can do omelettes without an egg, I normally get a series of cold calls from people selling me something to do with my energy bills. One day, I was experiencing the same manic Monday when I picked up the phone hoping it was going to be a reservation... instead it was someone calling about my energy usage. Just before I was about to put the phone down, the cold caller mentioned something about reducing my voltage. It was only because my electrician had said that I had very high voltage in the building the week before, that I drew breath and decided to listen...

We had 255 volts going through our building... so what did this mean? Well, most pieces of electrical equipment are set to work best with European standards of 200–220 volts, which meant that we were basically frying our equipment and decreasing its life expectancy. Also, we were using 55–35 volts per second more electricity than we needed, and therefore being charged 12–15% more on our bills than was necessary. We got a voltage optimizer and ended up saving £1,800 a year.

The point of all this? Well, it's going back to the question at start of the book: what is enough? Do you really need all that you are currently using? And are there ways of reducing what you use? It's a reminder to question this in everything you do.

Oh, and above all... try to install a voltage optimizer at work and/or at home.

EVERY ROOF NEEDS A BUTT

8

We wrongly tend to take it for granted, but fresh water will be considered one of the most valuable commodities of the future. As our population rises, there is constant demand on our basic resources – water, food and air. As we are all connected, whatever we waste contaminates our water, and so fresh water will be in scarcer supply.

One of the main aims of sustainability is relieving the pressure on our centralized industrial systems, which gives individuals more freedom and allows governments and states to spend more time and money managing other issues. At the moment it's absurd just how little water conservation we do on a local level. About 1% of all treated water in the UK ends up being drunk, and the rest is used for other things. That's a lot of time, energy and money spent on a process that isn't necessarily needed.

If every roof had a water butt, we could save huge amounts of clean rainwater to use for things other than drinking. Actually, as well as roofs, every terrace, patio and driveway could have a butt too! Just imagine how much water runs off these hard surfaces when it rains and bursts national water pipes. If you're landscaping a new garden, think about putting in a large rainwater harvesting tank below the ground.

A butt or tank will give you a ready supply of water for all your outdoor and indoor gardening desires. Rainwater is actually much better than tap water for all plants, including lawns, flowers, fruit and vegetables. This is because rainwater contains lots of lovely nutrients and avoids the chemicals and chlorine in tap water. Plants also benefit from rainwater being at a similar ambient temperature to the air around them, as opposed to tap water which is often much cooler.

A good filtration system means you could also use harvested rainwater for things like toilet flushing, clothes washing, household cleaning, showers and baths. Rainwater is also a naturally soft water, which helps counter limescale.

Grey water

This is the waste product left over from your baths, showers, sinks and washing machines. It is obviously not drinking water, but it can actually be treated and reused for things like washing clothes and flushing toilets. If you only use natural products, then whatever goes into your grey water is at least natural and therefore perfect for watering the garden. Grey water systems are simple to have installed in your home, especially in new homes. Another option is to remove the u-bend in the kitchen sink, place a bucket underneath to catch the water and use that water to flush your toilet.

Warm-up water

This is the water that we waste while we are waiting for the water to warm up from a tap or before we get in the shower. Catch it in a bucket and use it in the garden or around the house.

Other little things you can do to save water are reduce shower times, fix leaks and turn the tap off when brushing your teeth.

DECORATING

9

It's amazing the difference a woolly hat or thick socks can make to us on a cold day, so imagine what proper insulation and good double glazing can do to our homes. Insulation is the easiest way to reduce energy usage in the home, and it can be made from so many upcycled or recycled products. For example:

- Sheep's wool
- Recycled paper
- Recycled plastic bottles
- Plant fibre like hemp
- Sand and recycled glass
- Recycled corks
- Old newspapers
- Old unwearable clothes
- Old denim
- Old carpet tiles

Draught free

Draught-free homes need less heating because there's a steady controlled temperature rather than large fluctuations, so we don't need to crank the heating up high to balance out the cold draughts. To get rid of draughts, find covers for keyholes, letterboxes, flaps and door sides. Get some good insulation, block unused chimneys and fireplaces, fix draughty floorboards (or lay a large rug over them in winter).

Add some healthy colour

Who would have thought that paints could be toxic? The reality is that many of us are sleeping in rooms where chemicals and synthetic materials are leaching out of the walls all the time. If you don't want this, then use VOC-free paints. These paints are free from volatile organic compounds, which could be contributing to medical conditions like asthma and breathing difficulties.

There are also all sorts of other eco paints available, ranging from organic to vegan, plastic-free to metal-free. Either way, go for the most natural option and the best quality you can.

Lighting

Use LED lighting and bulbs. These are long lasting, emit little heat and are much more energy efficient than regular bulbs. Plus, they don't have the lead, mercury or other dangerous metals found in some alternatives. If you can, install motion sensor detectors for lighting in certain rooms in order to minimize the usage.

Recycle and upcycle

The most sustainable method of beautifying your home is to not create more demand on resources, but to reuse, recycle and upcycle what has already been made. Here are some inexpensive ideas for decorations:

- **Old apple crates** – use as bookshelves
- **Wooden ladders** – make nice bathroom shelves
- **Oyster shells and scallop shells** – make good salt holders
- **Hessian** – great for a table mat or some makeshift curtains
- **Wine bottles** – use as 80s-style candle holders
- **Old doors or palettes** – make cheap bed headboards
- **Large empty tins** – ideal for growing herbs or using as stationery/utensil holders
- **Dried flowers** – these are super easy to make and last for a few years. Simply bunch the fresh flowers together and leave them in a cool, dry, shaded spot inside. If you want them to remain as a bunch, it's best to leave them hanging upside down. Or, if you're going to hang them against a wall, then leave them to dry flat.

THE GARAGE

There is no doubt that the endless stream of cars up and down our motorways is one of the most obvious contributors to our carbon footprint. I really believe that technology will continue to improve in this area – we will become more creative in our processes and there will be more solutions to sustainable travel.

In the meantime, the first thing to note within the context of this book is that living sustainably means you are less likely to need to drive in the first place. Sourcing products for living from within our 30 miles will demand better local transport that involves sharing journeys.

Cars

The future of our planet surely has to depend on us switching to renewable energies. Therefore, electric cars are currently the best, but still highly expensive, option. However, the more public demand there is for electric cars, the more money will be invested in the companies who are developing the technology and rolling out the infrastructure. (Then there is the whole other question of lithium supplies, and how much we will need to extract in order for the whole world to have electric cars.)

Developments are constantly happening, but if you can't afford electric, then the next most sustainable option is to buy a second, third or fourth-hand car because it halts the demand for more resource extraction other than the fossil fuels to run the car. We should always be aiming to use unwanted resources first before extracting more.

Being mindful of the way we drive can help, too. Going at 50 mph (80 kmph) uses around 25% less fuel than driving at 70 mph (112 kmph). Accelerating and braking gently also helps.

And on the subject of transport, don't forget to consider the transport of the products we consume. The mass production of poor welfare beef, for example, is a huge contributor to carbon and methane emissions. You can change this immediately by altering how you shop (which is not quite as expensive as the electric car option).

Other ways to travel

Walking is always the best option for low carbon emissions, better individual health and thus reducing national health spending. It's free to use, improves air quality and reduces congestion. But when you need to go a bit further, consider using some these greener modes of transport:

- Push bikes
- Scooters
- Electric trains
- Electric buses
- Park and ride
- Car pool (there are some apps that allow people to travel in a journey that's already happening)

Working from home

Since the time of writing this book, Corona virus has completely changed our lives and many of us have had to work from home. Modern technology has shown us just how easy it can be to achieve this, and for many it has provided a more relaxed pace of life and an achievable work/life balance – our reduced travel has meant that the environment has reaped the benefits too.

Moving forward, if everyone did have one day a week working at home there would be 20% less cars on the road and therefore less pollution, easier and more efficient journeys and you would save yourself an average of two hours a week. Plus, you would add more to your local community by staying and perhaps buying your lunch in the area.

Carbon quotas

In around 10 years' time, I wouldn't be surprised if there was a carbon quota for each person. Each human on Earth will be allowed a certain amount of carbon per year, and we will have free reign over how to spend it. For example, if you want to take two long flights for a holiday one year, then you may not be able to drive for 20% of the same year. The same quotas might apply to businesses, and we could have carbon tax relief for greener companies. Economizing the one crucial resource that we need to reduce can only be a helpful thing.

Of course, quotas are a form of control, which is not ideal. But one of the aims of sustainability is to try to create more independent lives for us on a local level, so it's not the only way forward, but it is an option.

11 CHILDREN

When I imagine the best possible future for our children it is a sustainable one, because that's what creates the most delicious food, the healthiest people, provides the best economic model, the safest environment and leads to the most joyful lives.

Food is such a great tool to help young people understand the basic principles of sustainability and children should be exposed to raw, real foods from a young age. Some of my earliest memories include my mum shouting my name as I was curiously running after chickens, I was fascinated by this other life and amazed that they could produce eggs. I remember the realization that tomatoes came from my grandad's greenhouse and the smell of them that was even better than the taste. Or getting back from school and running to the garden to eat strawberries out of the pots, my white shirt sprayed with red. Memories are powerful and they stay with us forever. We should try to create the best memories we can for our children to act as strong foundations for their sustainable journeys ahead.

Sustainability in schools

If there are lessons about history in schools, then why can't we have lessons about the future? If the principles and importance of sustainability are instilled in young people from an early age, then all the Earth-positive habits I'm talking about in this book will become second nature. It could either be a subject by itself or included and applied across a variety of lessons. We need to radicalise our education systems so that when children leave school, they have a real understanding of the world outside and the part they play in it, allowing them to comprehend that whatever energy they create they will also consume.

Plastic not fantastic

The product market for children is chock-full of plastic. Throwaway plastic items might seem like the easiest, cheapest option for kids, especially when it comes to chucking out things that might be harbouring germs. But as we know, plastic items are actually the least healthy (as children will breathe in the microplastics), the most environmentally damaging and the most expensive option in the long run. There are plenty of hardier, healthier alternative materials to be used. Get rid of plastic lunchboxes, bottles and cutlery and instead use funky sustainable lunchboxes, wood and metal cutlery, metal bottles and cotton napkins. Of course use what you've already got first, but gradually learn how to avoid plastic with each necessary purchase you make. Instead of plastic toys, try giving children toys made out of wood, or better yet, reward them with paper books and colouring pencils or experiences, stories and make believe games instead.

Cloth nappies

An estimated 3 billion disposable nappies are thrown away in the UK each year, much of which will end up in landfill. The carbon footprint of cloth nappies compared to plastic ones is much lower impact, even more so if we use our water sustainably. Once you get used to them and get a good system going, cloth nappies can mean less prep for you, more comfort and less rashes for your little one. They are super easy to wash, organic, durable and long lasting, so can save you 50–80% on your nappy bills. Be sure to buy enough cloth nappies at the start so you don't quickly run out (about 20) and wash them regularly. (Also, always put poos in the loo, as human poo that goes to landfill can get into the water system).

There are times when cloth nappies might not be suitable, for example if you are on the road or travelling, so in these instances you could turn to disposable eco nappies instead, which are made from sustainable materials and free from chemicals. Currently, even eco nappies are not fully biodegradable though, so they should be considered a luxury rather than an essential.

Crispy Aubergine and Honey Bruschetta

Aubergines can be a tricky vegetable to get right, but once you know how to prepare them it will transform your veggie meals (and meaty ones) forever. One trick I use is to soak the aubergine in milk before cooking, which makes it really soft and delicious. Try to use a locally produced honey and this recipe will allow it to shine.

SERVES 4

1 AUBERGINE (EGGPLANT), THINLY SLICED INTO 0.5-CM/¼-INCH SLICES

100 ML/3⅓ FL OZ/⅓ CUP WHOLE MILK

2 SPRIGS OF FRESH THYME, LEAVES STRIPPED AND ROUGHLY CHOPPED

100 G/3½ OZ RUNNY HONEY

100 G/3½ OZ FLOUR (ANY TYPE WILL DO)

1 EGG, LIGHTLY WHISKED

100 G/3½ OZ/1¼ CUPS HOMEMADE DRIED BREADCRUMBS OR POLENTA (CORNMEAL)

LOCALLY PRODUCED OIL OF YOUR CHOICE WITH A HIGH SMOKING POINT, FOR SHALLOW-FRYING

4 SLICES OF SPELT BREAD

4 TBSP CRÈME FRAÎCHE

GOOD-QUALITY OLIVE OIL, FOR DRIZZLING

50 G/1¾ OZ COBNUTS OR OTHER LOCALLY GROWN NUTS, TOASTED AND ROUGHLY CHOPPED, TO SERVE

Put the aubergine slices in a large bowl, pour over the milk and season with salt then leave to soak at room temperature for 1 hour. Meanwhile, mix the thyme leaves with the honey and leave somewhere warm for the herb to infuse.

Put the flour, beaten egg and breadcrumbs or polenta each in separate shallow dishes. Take the aubergine slices out of the milk and shake off any excess liquid. Dip both sides of each soaked aubergine slice first in the flour, then in the egg, then in the breadcrumbs or polenta to give an even coating.

Fill a large, deep, heavy-based frying pan with 1¼ cm/½ inch of high-smoking point oil and place over a medium heat. When the oil is hot, add the aubergine slices and shallow-fry for 3 minutes on each side until crispy and golden. You may need to do this in batches depending on the size of your pan. Remove the aubergine slices from the oil with a slotted spoon and keep warm by covering or placing in a low oven.

Toast the slices of spelt bread and spread the toasts with one tablespoon each of crème fraîche. Divide the slices of crispy aubergine between the bruschetta, then drizzle over the thyme-infused honey followed by some good-quality olive oil. Scatter the bruschetta with toasted chopped nuts to serve.

TIP

You can use the milk leftover from soaking the aubergine in another recipe, such as a roux, soup or even to make homemade cheese.

Tomato Leaf Focaccia

If you are unsure as to whether tomato leaves are safe to eat, the important distinction to make here is that tomatoes are part of the nightshade, rather than the *deadly* nightshade, plant family. The stems, leaves and even immature unripened green tomatoes do contain solanine and tomatine, but the toxins are in such low levels that we would have to eat huge amounts for it to be harmful. (The real toxicity that we should be talking about is the amount of wasted food that we don't eat!) A classic chef's trick is to simmer tomato vines in soups for added flavour, and tomato leaves are best when chopped up and used like a herb.

MAKES 1 LOAF

10 TOMATO VINES PLUS LEAVES

500 ML/17 FL OZ/2 CUPS PLUS 2 TBSP HOT WATER

500 G/1 LB 2 OZ/SCANT 3⅔ CUPS STRONG BREAD FLOUR, PLUS EXTRA FOR DUSTING

1 TBSP SEA SALT, PLUS EXTRA FOR THE TOPPING

15 G/½ OZ FRESH YEAST OR 7 G/¼ OZ INSTANT DRIED YEAST

5 TBSP OLIVE OIL, PLUS EXTRA FOR DRIZZLING

HANDFUL OF SMALL TOMATOES (OPTIONAL)

2 GARLIC CLOVES, FINELY GRATED

Pick all the leaves off the tomato vines and set aside. Roughly chop the vines. Pour the hot water into a measuring jug (pitcher), add the chopped tomato vines and leave to infuse for 20 minutes.

Meanwhile, finely dice the tomato leaves. You should be left with about 40 g/1½ oz of chopped leaves. Add these to a large mixing bowl with the flour, salt, yeast and 2 tablespoons of the olive oil and mix well. Discard the tomato vines from the now lukewarm infused water and initially add 300 ml/10 fl oz/1¼ of water to the bowl. Mix with your hands to bring the dough together. Add some or all of the remaining water only if needed – it should be a soft dough but not too wet. Turn the dough out of the bowl onto a lightly floured work surface and knead by hand for 5 minutes until smooth and elastic. Lightly oil the same mixing bowl and return the dough to it. Cover with a plate or tea towel (kitchen cloth) and leave to prove in a warm place for about 1 hour or until doubled in size.

Turn the dough out onto a lightly floured work surface and knock it back by kneading it through again briefly. Roll the dough out to a rectangular shape, roughly the same size as a 30 x 20 cm/8 x 12-inch roasting pan with deep sides. Or, adjust the size of the dough to the size of pan you are using accordingly. Line the base of the pan with baking parchment, then place the rolled out dough in the pan and poke indentations all over it. (At this point you could add some small tomatoes if you like.)

In a small bowl, mix the grated garlic with the remaining 3 tablespoons of olive oil and some salt. Drizzle this evenly over the focaccia and spread evenly with your hands. Leave the dough to prove in a warm place for a final 20 minutes or until it has risen slightly again.

Preheat the oven to 220°C fan/240°C/475°F/gas mark 9.

Bake in the oven for 15 minutes or until evenly golden brown. Drizzle with a little more olive oil before eating warm. Focaccia is best eaten the same day it is made, but it will keep for up to 5 days and warming through will bring it back to life.

TIP

Use the leftover oil from some sun-dried tomatoes instead of olive oil for that extra tomato flavour and zero waste positivity. If you've made the sun-dried tomatoes yourself then even better!

Spelt, Seaweed and Rosemary Bread

Dried seaweed is very easy to use and adds a naughty saltiness to dishes. If you're lucky enough to live by the sea, then fresh seaweed is even better. Springtime or cooler seasons are the best times to go foraging on the beach, and it makes for a really pleasant walk too. There are lots of varieties, so do your research and make sure you identify correctly before picking or eating. If you can use scissors to cut off a piece of seaweed, it should mean that the rest of the plant left behind can still grow.

MAKES 1 LOAF

40 G/1½ OZ DRIED SEAWEED FLAKES OR 80 G/2¾ OZ FRESH SEAWEED, FINELY CHOPPED

360 ML/12 FL OZ/1½ CUPS LUKEWARM WATER

500 G/1 LB 2 OZ/GENEROUS 3¾ CUPS WHOLEGRAIN SPELT FLOUR, PLUS EXTRA FOR DUSTING

50 G/1¾ OZ/⅓ CUP PUMPKIN SEEDS

1 TBSP SEA SALT

1 TBSP GOLDEN CASTER (GRANULATED) SUGAR

15 G/½ OZ FRESH YEAST OR 7 G/¼ OZ INSTANT DRIED YEAST

3 SPRIGS OF FRESH ROSEMARY, LEAVES FINELY CHOPPED

If using dried seaweed flakes, place them in a large mixing bowl, pour over the lukewarm water and leave for 5 minutes. If using fresh seaweed, do the same but there's no need to wait for 5 minutes.

Add all the remaining ingredients to the bowl and stir thoroughly to bring together into a dough. Turn the dough out onto a lightly floured work surface and knead through by hand for 10 minutes until smooth and elastic. Lightly oil the same mixing bowl and return the dough to it. Cover with a plate or tea towel (kitchen cloth) and leave the dough to prove in a warm place for about 1 hour or until doubled in size.

Lightly oil a 1 kg/2 lb 3 oz loaf pan. Turn the dough out onto a lightly floured work surface and knead through again for 3–5 minutes. Shape into a loaf and transfer to the oiled loaf pan. Cover the dough with a tea towel (kitchen cloth) and leave to prove in warm place for a final 30 minutes or until doubled in size again.

Preheat the oven to 200°C fan/220°C/425°F/gas mark 7.

Bake for 30 minutes until risen and golden and the loaf sounds hollow when tapped on the base. Leave the loaf to cool on a wire rack before slicing.

Camelina Seed Crackers

These crackers are the easiest little delights served with any cheese and a good dollop of chutney. Camelina has been grown in England for centuries but has lost some of the limelight in recent years. There are many such ancient grains and seeds like this that have been overlooked in favour of the latest trends. You can help keep the diversity in our ecosystems by mixing it up and using lesser-known produce from time to time.

MAKES ABOUT 30 CRACKERS

150 G/5⅓ OZ/SCANT 1¼ CUPS WHOLEGRAIN SPELT FLOUR, PLUS EXTRA FOR DUSTING

2 TSP LOCALLY PRODUCED OIL OF YOUR CHOICE

7 G/ ¼ OZ SEA SALT

40 G/1½ OZ CAMELINA SEEDS (FLAXSEEDS/ LINSEEDS ALSO WORK WELL HERE)

Preheat the oven to 200°C fan/220°C/425°F/gas mark 7.

Combine all the ingredients in a mixing bowl with 75 ml/2½ fl oz water and bring together into a dough with your hands. Tip the dough out onto a lightly floured work surface and knead through until the ingredients are well mixed and the dough is smooth.

Cut the dough in half, cover one half with a tea towel (kitchen cloth) and roll the other half out until it's as thin as the seeds – I really mean as thin as the seeds. Cut into shapes using a knife or stamp cutter. You can get creative here or just do simple rectangles and make sure you re-roll and cut out any scraps. Transfer the dough shapes to a lightly oiled baking sheet and repeat with the other half of dough, transferring the cut shapes to a second oiled baking sheet.

Bake the crackers for 5 minutes until the dough is cooked but not browned. Transfer to a wire rack to cool before eating or storing. The crackers will keep in an airtight container for up to 1 month.

A NOTE ON STICHELTON AND OTHER LOCAL CHEESES

If you haven't heard of or eaten Stichelton, then you're in for a real treat – and yes, it is spelled that way. They weren't allowed to call it a Stilton, so they invented a name... I love the cheese, the story and the name. It's a traditional cheese made with raw, organic milk in Nottinghamshire. I believe it can compete with the best cheeses in the world. It just proves that we don't need to be buying ridiculous amounts of cheese from abroad. Stichelton isn't within my 30 miles, but it is a rare treat and at least produced in my home country. I recommend finding the best cheeses that are produced near you and singing their praises when you find a good one.

Coffee Grounds Bread

This is ultimately why I love sustainability - because it really produces the most delicious food. At the pub we end up with a lot of leftover coffee grounds, so I had to find a way of using them up. We had previously made a cracker using coffee beans, pairing it with Comté cheese and honey, so I thought perhaps it would work in a bread to go with our Welsh rarebit. We made it and it worked brilliantly, with a flavour somewhere between stout, rye and malt. It makes for a nutty, slightly bitter, coffee-scented loaf that brings flavour satisfaction and planet-loving thrills.

MAKES 1 LOAF

500 G/1 LB 2 OZ/GENEROUS 3¾ CUPS WHOLEGRAIN SPELT FLOUR, PLUS EXTRA FOR DUSTING

15 G/½ OZ SEA SALT

40 G/1½ OZ/SCANT ¼ CUP SOFT DARK BROWN SUGAR

150 G/5⅓ OZ USED COFFEE GROUNDS

15 G/½ OZ FRESH YEAST OR 7 G/¼ OZ DRIED INSTANT YEAST

300 ML/10 FL OZ/1¼ CUPS WARM WATER (AROUND BODY TEMPERATURE - 38°C/100°F)

Put the flour, salt, sugar and coffee grounds in a large bowl and mix well. Put the yeast and 270 ml/9 fl oz of the warm water in a separate bowl and mix until dissolved. Add the yeast water to the bowl with the flour. Mix to bring together into a dough, adding the remaining warm water only if needed to bring it together. Turn the dough out onto a lightly floured work surface and knead for 8 minutes by hand until it is smooth and elastic.

Lightly oil the same mixing bowl and return the dough to it. Cover with a plate or tea towel (kitchen cloth) and leave to prove in a warm place for about 1 hour or until doubled in size.

Lightly oil a 900 g/2 lb loaf pan. Turn the dough out onto a lightly floured work surface and knock it back by kneading through again briefly. Shape into a loaf and transfer to the oiled loaf pan. Cover with a tea towel (kitchen cloth) and leave to prove in a warm place for a final 40 minutes or until the dough has risen above the top of the pan.

Preheat the oven to 200°C fan/220°C/425°F/gas mark 7.

Bake for 30 minutes until risen and golden and the loaf sounds hollow when tapped on the base. Leave to cool on a wire rack before slicing.

Dulse and Oat Croquettes

A meat-free take on the classic Spanish snack of jamón croquettes.

MAKES ABOUT 15

FOR THE FILLING

65 G/2¼ OZ/⅔ STICK BUTTER

3 TBSP LOCALLY PRODUCED OIL OF YOUR CHOICE

1 ONION, FINELY DICED

6 GARLIC CLOVES, FINELY DICED

BUNCH OF PARSLEY, STALKS AND LEAVES SEPARATED AND DICED

3 BAY LEAVES

65 G/2¼ OZ/½ CUP PLAIN (ALL-PURPOSE) FLOUR

100 G/3½ OZ/1 CUP ROLLED (OLD FASHIONED) OATS

40 G/1½ OZ DRIED DULSE SEAWEED FLAKES, ROUGHLY CHOPPED

500 ML/17 FL OZ/2 CUPS PLUS 2 TBSP WHOLE MILK

A LITTLE FRESHLY GRATED NUTMEG

LOCALLY PRODUCED OIL OF YOUR CHOICE WITH A HIGH SMOKING POINT, FOR DEEP-FRYING

FOR THE COATING

2 EGGS, BEATEN

50 G/1¾ OZ/GENEROUS ⅓ CUP PLAIN (ALL-PURPOSE) FLOUR

100 G/3½ OZ/1¼ CUPS HOMEMADE DRIED BREADCRUMBS

For the filling, put the butter and oil in a saucepan over a medium heat. Add the onion, garlic, parsley stalks and bay leaves and cook for 5 minutes, stirring. Add the flour, oats and dulse and cook for 5 more minutes, stirring. Turn the heat to low, then add the milk in batches, mixing well, until you have a smooth white sauce. Cook for 15 minutes to thicken. Stir in the parsley leaves and nutmeg, then cool and refrigerate the sauce for at least 2 hours until firm.

Put the eggs, flour and breadcrumbs into separate dishes. Roll spoonfuls of the white sauce into 15 balls, then roll each one in the flour, then the egg, then the breadcrumbs.

Preheat a deep fat fryer or half-fill a heavy-based pan with oil and bring the temperature to 180°C (356°F). Deep-fry the croquettes in batches for 2 minutes until golden. Remove with a slotted spoon and serve immediately.

Celeriac Top and Apple Soup

Celeriac tops have a similar taste to celery and the young stalks and leaves can be used in the same way. But with celeriac harvested at its peak, the tops have grown quite tall and thick and can be a bit stringy. I find it's best to slurp them down in this autumnal, flavour-rich but light soup, to which apples add fantastic texture.

SERVES 4

LOCALLY PRODUCED OIL OF YOUR CHOICE, FOR FRYING

1 ONION, DICED

1 FENNEL BULB, DICED

250 G/9 OZ CELERIAC TOPS (ABOUT 2 CELERIAC HEADS), DICED

200 G/7 OZ PEELED, CORED AND ROUGHLY CHOPPED APPLE FLESH (ABOUT 300 G/10½ OZ APPLES)

SEA SALT

Add a few glugs of oil to a large saucepan and place over a low–medium heat. Add the diced onion and fennel and cook for about 6 minutes, stirring occasionally, until soft.

Add 1 litre/quart of water, bring to the boil, then turn down the heat and simmer for 10 minutes.

Add the diced celeriac tops and apple flesh to the pan and simmer for a further 5 minutes. Transfer to a food processor and blitz until smooth. Season with salt and warm through in the pan again to serve.

TIP

Combine the apple cores and peels from this recipe with 1 cinnamon stick, 3 cloves and 1 litre/quart of vinegar. Leave to infuse to make a scented cleaning product!

Dandelion and Courgette Pakoras with Spiced Beetroot Yogurt

Dandelions are best eaten in the spring, when their leaves are young and tender. As we all know, they grow abundantly and they're super scrummy as well as being a perfect root-to-fruit plant packed full of nutrients. They can be used in all sorts of things, from salads to stews. Their root even makes a wonderful brew.

MAKES 12 PAKORAS

FOR THE PAKORAS

LARGE BUNCH OF FRESH DANDELION LEAVES

2 ONIONS, THINLY SLICED

3 COURGETTES (ZUCCHINI), GRATED

12 TBSP BROAD (FAVA) BEAN FLOUR OR FLOUR OF YOUR CHOICE

1 TSP CUMIN SEEDS

1 TSP SEA SALT

1 TSP GROUND TURMERIC

1 TSP GROUND CORIANDER

1 TSP GROUND GINGER

½ TSP DRIED CHILLI (HOT RED PEPPER) FLAKES

LOCALLY PRODUCED OIL OF YOUR CHOICE WITH A HIGH SMOKING POINT, FOR DEEP-FRYING

FOR THE SPICED BEETROOT YOGURT

2 RAW BEETROOTS

6 TBSP HOMEMADE YOGURT (SEE PAGE 58)

SMALL BUNCH OF FRESH MINT, DESTALKED

1 TSP GROUND CUMIN

2 TBSP LOCALLY PRODUCED VINEGAR OF YOUR CHOICE

100 ML/3⅓ FL OZ/⅓ CUP LOCALLY PRODUCED OIL OF YOUR CHOICE

For the pakoras, soak the dandelion leaves in a bowl of cold water for 3–4 hours to remove some of the bitterness.

Meanwhile, make the spiced beetroot yogurt. Preheat the oven to 180°C fan/200°C/400°F/gas mark 6.

Place the beetroots in a roasting pan and roast in the oven for 45 minutes or until soft when pierced with a knife. Leave to cool, then remove and discard the skins and roughly chop the flesh. Place the roasted beetroot in a food processor with the yogurt, mint, cumin, vinegar and oil. Blitz until smooth, then set aside until needed.

When the dandelion leaves have finished soaking, drain the water and dry them, then finely dice them up. Place in a mixing bowl with the remaining pakora ingredients. The water in the courgettes should bring the mixture together to form a thick batter, but if not then add a few tablespoons of water.

Preheat a deep-fat fryer to 170°C (340°F) or half-fill a deep, heavy-based pan with oil, place over a medium heat and bring the temperature to 170°C (340°F). Pick up a heaped tablespoonful of pakora batter and carefully drop it into the hot oil. Add more spoonfuls of batter (how many depends on the size of your pan, just don't overcrowd it) and deep-fry for about 4 minutes until golden and crispy. Remove with a slotted spoon and briefly drain the excess oil on kitchen paper or a tea towel (kitchen cloth). Keep the pakoras warm in a low oven while you fry the rest of the batter.

Serve the hot pakoras with the spiced beetroot yogurt.

Brussels Sprout Tops and Sea Greens with Fermented Broad Bean Broth

You don't really think about what ingredients like soy sauce and miso are made from until someone tells you. If someone had told me that soy sauce is made from fermented soy beans when I was growing up, I probably would have thought twice before lashing it over my noodles. I've since learned not to let a weird sounding name put me off. When one of my suppliers at the pub told me to try their fermented broad bean umami paste, I was intrigued. I ordered a pot to try and was blown away. It's real confirmation that we can achieve culinary greatness by utilizing local produce from within our own 30-mile radius - we don't have to look to the other side of the world for exciting produce.

This broad bean paste is so deeply rich and satisfying that it actually took me a while to work out what to do with it. I was experimenting on another dish using Brussels sprout tops and sea greens when my cousin, Rosa Rigby, suggested using the paste like a miso. Bang… it worked so well. Thank you, Rosa, for this wonderful idea - you flavour jedi (Obi would be proud).

SERVES 4

2 SMALL TBSP FERMENTED BROAD (FAVA) BEAN PASTE OR ANOTHER LOCALLY PRODUCED FERMENTED BEAN PASTE, PLUS EXTRA FOR ROASTING THE SEEDS

4 TBSP PUMPKIN SEEDS

LOCALLY PRODUCED OIL OF YOUR CHOICE WITH A HIGH SMOKING POINT, FOR FRYING

4 GARLIC CLOVES, THINLY SLICED

2 RED CHILLIES, THINLY SLICED

5-CM/2-INCH PIECE OF FRESH GINGER, THINLY SLICED

8 BROWN MUSHROOMS, THINLY SLICED

SMALL BUNCH OF FRESH CORIANDER (CILANTRO), STALKS AND LEAVES SEPARATED AND STALKS THINLY SLICED

15 G/¾ OZ DRIED SEA GREENS

150 G/5⅓ OZ BRUSSELS SPROUT TOPS, BOTTOM CENTRAL RIB REMOVED AND THINLY SLICED

200 G/7 OZ RICE NOODLES OR HOMEMADE LINGUINE (SEE PAGE 94)

FRESHLY SQUEEZED JUICE OF 1 LIME, TO SERVE

Preheat the oven to 180°C fan/200°C/400°F/gas mark 6.

Place the pumpkin seeds on a baking sheet and rub with a little of the fermented broad bean paste (enough to lightly coat them). Roast in the oven for 4 minutes, then remove and leave to cool.

Place a large saucepan over a high heat with a few glugs of locally produced oil. Add the sliced garlic, chillies, ginger, mushrooms and coriander stalks. Stir-fry for 3–4 minutes. Add the dried sea greens and stir-fry for a further minute. Spoon in the fermented bean paste and stir-fry for 30 seconds, then add 1.5 litre/1 ½ quart/ 6 cups of water. Bring to a simmer, then add the Brussels sprout tops and noodles or pasta for the final 5 minutes of cooking.

Turn off the heat and finish with a squeeze of lime juice. Serve in bowls scattered with the coriander leaves and the roasted pumpkin seeds.

Soul warming and planet cooling.

Baba Ganoush with Redcurrants, Mint and Coriander

For me, baba ganoush has to be smoky. A little charring of the aubergine either on the barbecue, over a gas hob (stove) or - if you have to - under the grill (broiler) transforms this dish to new heights. It's a dip that is normally served with pomegranates for that classic Persian flavour, but here I'm using redcurrants instead as they still give that pop of sweet red juice without the air miles. Likewise, for the same reason I've also substituted classic sesame tahini for blended sunflower seed tahini, but the choice is yours.

SERVES 4 AS AN APPETIZER OR SIDE DISH

3 MEDIUM AUBERGINES (EGGPLANT)

2 GARLIC CLOVES, GRATED

2 TBSP SUNFLOWER SEEDS, ROASTED IN THE OVEN FOR 5 MINUTES

1 RED CHILLI, ROUGHLY CHOPPED

1 TSP GROUND CUMIN

1 TSP GROUND CORIANDER

FRESHLY SQUEEZED JUICE OF 1 LEMON

SALT

LOCALLY PRODUCED OIL OF YOUR CHOICE, FOR DRIZZLING

TO SERVE

100 G/3½ OZ/1 CUP FRESH REDCURRANTS

SMALL BUNCH OF FRESH MINT, DESTALKED AND LEAVES ROUGHLY CHOPPED

SMALL BUNCH OF FRESH CORIANDER (CILANTRO), LEAVES ONLY, ROUGHLY CHOPPED (STALKS KEPT FOR ANOTHER DISH)

If using a barbecue (grill) then get it to a low steady heat and add the aubergines. Let them slowly cook for about 10 minutes, turning every now and again or until the skin is crisp and the flesh is gooey and soft. If you're cooking over a gas hob (stove top), turn the flame to low-medium, place a wire rack over the heat and cook for about 6 minutes in the same way. If using a grill (broiler), then preheat to medium and cook the aubergines for about 6 minutes in the same way.

Once they're cooked, put the aubergines in a large bowl and immediately cover with a plate. Leave for 15 minutes and the steam should make the skins very easy to peel away. Peel and discard all the skins and put all the gooey flesh in a food processor with the garlic, roasted sunflower seeds, chilli, cumin, coriander, lemon juice and a bit of salt. Blitz to a paste and then slowly drizzle in oil while blending until the baba ganoush is smooth and about as thick as hummus.

Place your baba ganoush in a serving bowl and top with extra drizzles of oil, scatterings of redcurrants, chopped mint and coriander.

Barbecued Leeks with Cheddar, Poached Eggs and Tarragon

Barbecuing is a cooking method that really brings out our primal instincts, adding the smoky flavour of fire that we have known for thousands of years. This delicious cooking method shouldn't just be reserved for meat though, it can be an important tool in our sustainable mission to eat more vegetables. Proper fresh, organic baby leeks cooked on a gentle barbecue are a thing of beauty and, just like all baby vegetables, they have a natural sweetness that is second to none. Just add an oozy egg with cheese and tarragon butter to seal the deal. This dish is the perfect start to a summer evening.

SERVES 4 AS A LIGHT MEAL OR APPETIZER

12 YOUNG FRESH BABY LEEKS

3 TBSP LOCALLY PRODUCED VINEGAR OF YOUR CHOICE

LOCALLY PRODUCED OIL OF YOUR CHOICE, FOR DRIZZLING

SMALL BUNCH OF FRESH TARRAGON, FINELY CHOPPED

VERY SMALL BUNCH OF FRESH PARSLEY, FINELY CHOPPED

150 G/5⅓ OZ/1¼ STICKS ROOM TEMPERATURE BUTTER

4 EGGS (IDEALLY AS FRESH AS POSSIBLE)

150 G/5⅓ OZ CHEDDAR OR ANOTHER LOCALLY PRODUCED HARD CHEESE

SEA SALT

A gentle slow cook is the best way to ensure the insides of your leeks are cooked as well as the outsides. Therefore, preheat your barbecue (grill) to a temperature where you could carefully leave your hand above the rack for 10 seconds, but no more. As in it's definitely hot, but not scorching hot. Alternatively, if not using a barbecue, turn the grill (broiler) onto a medium-low heat and follow the method as below.

Cut off the very top green parts of the leeks, reserving them for another dish. Slice the remaining leeks in half lengthways, 5 cm/2 inches from the top (so the green parts open out slightly and the white ends stay intact). Clean the leeks by rinsing them and then tipping upside down to ensure any dirt falls out rather than embeds itself deeper.

Half-fill a saucepan with water and add the vinegar. Bring to a simmer ready to poach your eggs.

Drizzle the leeks with a little oil, season with salt and then put on the barbecue rack. Cook for 5–10 minutes, turning occasionally.

Meanwhile, mix the chopped herbs with the butter. When the leeks are completely soft and slightly charred, transfer them to a plate and spoon most of the herby butter all over the top.

Crack the eggs into the simmering water to poach for 3 minutes. Lift out the eggs with a slotted spoon and then place them on top of the leeks.

Spoon the rest of the herby butter over the eggs and then grate over the hard cheese to finish. Devour.

Chickpea Scotch Eggs with Aquafaba Aioli

Something that has changed for the better since I wrote my first book is that chickpeas are now commercially grown in England. This recipe is a delicious way to use them, along with their aquafaba by-product. You can try other pulses like broad (fava) beans in scotch egg form too, or whatever is grown locally to you.

MAKES 4 SCOTCH EGGS AND ABOUT 500 ML/17 FL OZ AIOLI

FOR THE SCOTCH EGG FILLING

4 EGGS

1 X 400-G/14-OZ CAN COOKED CHICKPEAS

3 TBSP SUNFLOWER SEEDS

1 SMALL RED ONION, FINELY CHOPPED

3 TBSP CHOPPED FRESH SAGE

1 TBSP CHOPPED FRESH THYME

A LITTLE FRESHLY GRATED NUTMEG

1 TBSP ANY TYPE OF FLOUR

SEA SALT

TO FINISH THE SCOTCH EGGS

60 G/2 OZ ANY TYPE OF FLOUR

100 G/3½ OZ/1¼ CUPS HOMEMADE DRIED BREADCRUMBS OR POLENTA (CORNMEAL)

OIL OF YOUR CHOICE WITH A HIGH SMOKING POINT, FOR DEEP-FRYING (OPTIONAL)

FOR THE AQUAFABA AIOLI

120 ML/4 FL OZ/½ CUP AQUAFABA FROM THE CHICKPEA CAN

2-3 GARLIC CLOVES TO TASTE, GRATED

1 TSP MUSTARD (LOCAL, YOUR CHOICE)

2 TBSP LOCALLY PRODUCED VINEGAR

450 ML/15 FL OZ/2 CUPS COLD-PRESSED OIL

FRESHLY SQUEEZED JUICE OF ½ LEMON

Cook the eggs in a saucepan of simmering water for 8 minutes until hard-boiled, then drain and immediately dunk them into ice-cold water. Leave for 5 minutes, then peel and set aside.

To make the scotch egg filling, drain the aquafaba from the can of chickpeas into a bowl and set aside. Place the chickpeas in a food processor. Toast the sunflower seeds for a few minutes in a dry frying pan (skillet), then tip into the food processor. Add the red onion, sage, thyme, nutmeg, flour and a good pinch of salt to the food processor. Blend together to make a smooth paste. Set aside for a moment.

To make the aioli, add 120 ml/4 fl oz/½ cup of the aquafaba, the grated garlic, mustard and vinegar to a bowl or container with high sides and blitz to combine with a hand-held stick blender. You can do this in a food processor if you prefer. Then, as with regular mayo, slowly pour in the oil while blending. It should emulsify nicely. Mix in the lemon juice and some salt to finish.

To assemble the scotch eggs, mix the remaining aquafaba with 2 tablespoons of cold water in a shallow dish and set aside. Place the flour and breadcrumbs on separate plates. Roll each cooked egg in flour to coat all over, then wrap a quarter of the chickpea mixture around each egg. Roll each egg in the flour again, then coat in the aquafaba mixture, then finally roll in the plate of breadcrumbs to evenly coat all over.

Preheat a deep fat fryer to 170°C (340°F) or half-fill a deep, heavy-based pan with oil, place over a high heat and bring the temperature to 170°C (340°F). Deep-fry the scotch eggs in two batches for about 5 minutes until golden and crispy. Remove with a slotted spoon and briefly drain the excess oil. Alternatively, you could roast the scotch eggs at 180°C fan/200°C/400°F/gas mark 6 for 8 minutes.

Serve with plenty of your homemade aioli for dipping.

TIP For a lighter, plant-based replacement for beurre blanc or hollandaise, mix two-thirds of the mayo recipe (minus the garlic) with one-third water and heat through very gently.

Homemade Yogurt

Who would have thought that making yogurt could be this easy! A cook's thermometer helps here, but if you don't have one then it's still very simple. The great thing about making yogurt yourself, other than saving on the horrific amount of single-use plastics, is the probiotic properties it contains - the yeasts are really great for your digestive system. Plus, the more sustainable and better quality the milk, the better your yogurt will taste.

MAKES ABOUT 600 G/1 LB 5 OZ YOGURT

570 ML/19 FL OZ/SCANT 2½ CUPS WHOLE MILK

2 TBSP ROOM TEMPERATURE LIVE YOGURT OR A YOGURT STARTER OR 12 CHILLI STALKS (NOTE THAT THE YOGURT WILL HAVE A SLIGHT CHILLI TASTE IF USING CHILLI STALKS)

Pour the milk into a saucepan and bring to the boil, then leave to cool in the pan until the temperature measures 40–45°C (104–113°F) on a thermometer, or until you can't touch it for more than 3 seconds.

Meanwhile, pour some hot water into the container you're about to use and leave it for 5 minutes, this will help maintain the temperature.

If using live yogurt, then make sure it is room temperature (if it isn't then you can mix in 1 tablespoon of the hot milk at a time to raise its temperature).

Pour away the hot water from the container. Add the live yogurt or yogurt starter or chilli stalks to the milk in the pan and stir. Pour the mixture into your warmed container and cover with a lid. Wrap the whole container up tightly in two tea towels (kitchen cloths) to keep the heat in and leave somewhere draught-free and on the warm side of room temperature for 6–8 hours. When you come back, you should find it naturally set.

Transfer the container of yogurt to the fridge and leave to chill for at least an hour before using. It will keep in the fridge for up to 2 weeks.

TIPS FOR SUCCESS

- Always use a sterilized container for each batch.
- The milk/yogurt mixture needs to maintain its heat for a few hours, so check it regularly and move it to a warmer place if needed or wrap in more towels.
- If you want a sweeter yogurt, you can simmer the milk for 5–10 minutes after boiling to create some of those natural sugars.
- If your yogurt is too runny after 6–8 hours, it means that the initial milk temperature was too low or not enough live yogurt/starter/chilli was added. In this case – place in a pan, bring to the boil, then turn off the heat and let it curdle. Strain to separate the curds and whey. Use the whey in other products like soda bread. The curds are delicious on toasted soda bread.
- Your yogurt will turn sour if the milk was too hot or too much live yogurt/starter/chilli was added. You can try to fix it by mixing in 100 ml/3⅓ fl oz water, leaving for 10 minutes and then straining. If it doesn't help then sour yogurt is good as a hair conditioner or used as a face mask.

Lentil Bolognese with Herby Pangrattato

This satisfying dish is the edible equivalent of a comforting hug, and because we have swapped the usual meat for veg and used locally grown lentils, it brings that sustainable feel-good factor too. I love to use this ragù in a myriad of ways, whether with pasta, polenta, on a baked potato or to stuff a giant squash with. The mushrooms and aubergine both work perfectly to give a bit of umami.

SERVES 4

FOR THE BOLOGNESE

LOCALLY PRODUCED OIL OF YOUR CHOICE, FOR FRYING

2 ONIONS, FINELY DICED

6 GARLIC CLOVES, FINELY DICED

2 CARROTS, FINELY DICED

2 CELERY STALKS OR YOUNG CELERIAC STALKS, FINELY DICED

10 MUSHROOMS OF YOUR CHOICE, DICED

1 AUBERGINE (EGGPLANT), FINELY DICED

BUNCH OF FRESH ROSEMARY, CHOPPED

BUNCH OF FRESH OREGANO, CHOPPED

150 ML/5 FL OZ/⅔ CUP RED WINE

400 G/14 OZ LOCALLY GROWN DRIED LENTILS

2 X 400 G/14-OZ CANS OF CHOPPED TOMATOES

2 BAY LEAVES

SEA SALT

LOCAL HARD CHEESE, TO SERVE

FOR THE PANGRATTATO

HANDFUL OF FRESH ROSEMARY, CHOPPED

HANDFUL OF FRESH SAGE, CHOPPED

HANDFUL OF FRESH THYME, CHOPPED

HANDFUL OF FRESH OREGANO, CHOPPED

100 G/3½ OZ/1¼ CUPS HOMEMADE DRIED BREADCRUMBS

DRIZZLE OF LOCALLY PRODUCED OIL

Add a few good glugs of oil to a large saucepan and place over a low-medium heat. Add the diced onions, garlic, carrots, celery (or celeriac), mushrooms and aubergine cook slowly for about 15 minutes. Add the chopped rosemary and oregano to the pan, stir and cook for a further 5 minutes.

Turn the heat up to high, add the wine and let bubble for about 5 minutes or until reduced by half. Add the lentils, chopped tomatoes, bay leaves and some salt and mix together. Pour 1 litre/quart of water into the pan, bring to a simmer and cook, uncovered, for about 45 minutes until the lentils are soft.

Meanwhile, preheat the oven to 180°C fan/200°C/400°F/gas mark 6.

Add all the finely chopped herbs for the pangrattato to a baking sheet with the breadcrumbs and oil. Season with a pinch of salt and mix thoroughly. Bake for 8 minutes, then remove and give the breadcrumbs a stir, then cook for a further 5 minutes until just golden.

Serve the bolognese with something starchy like crispy polenta or pasta or on a jacket potato. Grate over plenty of hard cheese and scatter with the herby pangrattato to finish.

Pea Pancakes with Purple Sprouting Broccoli, Squash and Yogurt

Making use of a diverse range of ingredients is a great way to cook sustainably. It brings a myriad of flavours to our kitchens, as well as a plethora of nutrients for the soil. Rather than just deferring to plain wheat flour all the time, one of my favourite things to do is come up with new ways of using different types of flour in cooking... potato flour gnocchi, broad (fava) bean flour bhaji batter, quinoa flour soufflés, buckwheat flour pizza base, acorn flour sponge cakes... the possibilities are endless.

I'm using pea flour here, which gives a creamy flavour and a punch of protein to these pancakes. The pancakes would work well for breakfast served with an egg and some kippers, or perhaps with harissa, feta and roasted nuts. Or maybe use the batter for coating fried fish and chips?!

SERVES 4 (MAKES 4 LARGE PANCAKES)

FOR THE BROCCOLI AND SQUASH

1 SQUASH (I LOVE USING UCHIKI KURI)

2 GARLIC CLOVES, FINELY CHOPPED

1 RED CHILLI, FINELY CHOPPED

2 SPRIGS OF FRESH ROSEMARY, LEAVES FINELY CHOPPED

1 TBSP FENNEL SEEDS

400 G/14 OZ PURPLE SPROUTING BROCCOLI, TRIMMED AND ANY LARGER STALKS SPLIT IN HALF

LOCALLY PRODUCED OIL OF YOUR CHOICE, FOR ROASTING

SEA SALT

FOR THE PANCAKES

150 G/5⅓ OZ/SCANT 1¼ CUPS PEA FLOUR (OR USE ANY FLOUR YOU LIKE AND ADJUST THE QUANTITY OF WATER ACCORDINGLY)

1 TSP CHOPPED FRESH MINT LEAVES

LOCALLY PRODUCED OIL OF YOUR CHOICE WITH A HIGH SMOKING POINT, FOR FRYING

TO SERVE

HOMEMADE YOGURT (SEE PAGE 58)

8 FRESH CHIVES, CHOPPED

CHIVE FLOWERS (OPTIONAL)

Preheat the oven to 200°C fan/220°C/425°F/gas mark 7.

Cut the squash in half, remove and discard the stringy insides and reserve the seeds to one side. Leave the skin on or peel your squash, depending on whether the skin is soft enough to eat on the variety you are using, and chop into lengths of a similar size and length to the purple sprouting broccoli. Transfer the squash to a large roasting pan with the garlic, chilli, rosemary and fennel seeds. Add a few glugs of oil and season with salt. Mix well, then roast for 10 minutes.

Meanwhile, mix together the pea flour, chopped mint and a generous pinch of salt in a mixing bowl with 200–250 ml/6¾ fl oz/generous ¾ cup water to make a batter. Leave it to rest for 10 minutes.

While the batter is resting, drizzle the purple sprouting broccoli with a little oil and season with salt, then add to the roasting pan along with the squash seeds. Give everything a good mix, then continue roasting for a further 10 minutes.

Place a large frying pan (skillet) over a medium heat with a good glug of oil. Add 2–3 large spoonfuls of the pancake batter to the pan and immediately tilt the pan to evenly distribute the mixture to the edges. Cook for 3–5 minutes until the bottom of the pancake is golden, then flip and cook for a further 3 minutes until golden on both sides. Remove to a plate and keep warm while you cook the remaining batter.

Place a pea pancake on each serving plate, dollop over some yogurt and then pile a helping of roasted squash and broccoli on top. Garnish with some chopped chives and chive flowers, if you like.

Broad Bean Burgers with Goat's Cheese, Apple Glaze and Pickled Beetroot

This tasty veggie burger hits the spot with its many layers of flavour. The pickled beetroot makes more than you need but it will keep for at least a month.

MAKES 4 BURGERS

FOR THE PICKLED BEETROOT

100 ML/3⅓ FL OZ/⅓ CUP RED WINE VINEGAR

200 ML/6¾ FL OZ/¾ CUP CIDER VINEGAR

100 G/3½ OZ/½ CUP GOLDEN CASTER (GRANULATED) SUGAR

1 STAR ANISE

10 CORIANDER SEEDS

10 CUMIN SEEDS

5 FENNEL SEEDS

2 RAW BEETROOTS

FOR THE APPLE GLAZE

1 LITRE/QUART/4 CUPS APPLE JUICE

280 ML/9½ FL OZ/SCANT 1¼ CUPS APPLE CIDER VINEGAR

120 G/4¼ OZ/SCANT ⅔ CUP GOLDEN CASTER (GRANULATED) SUGAR

FOR THE BURGERS

600 G/1 LB 5 OZ SWEET POTATOES

240 G/8½ OZ COOKED FAVA (BROAD) BEANS (1 X 400-G/14-OZ DRAINED CAN)

2 TSP GROUND CUMIN

2 TSP GROUND CORIANDER

2 TSP GROUND DRIED CHILLI (HOT RED PEPPER) FLAKES

70 G/2½ OZ/½ CUP BROAD (FAVA) BEAN FLOUR

TO SERVE

4 SLICES OF GOAT'S CHEESE

BRIOCHE BURGER BUNS (SEE PAGE 95)

CONDIMENTS OF YOUR CHOICE

Start by making the pickled beetroot. Stir together both vinegars, the sugar and all the spices in a saucepan and bring to the boil for 3 minutes, then remove from the heat. Meanwhile, peel the beetroots, then continue using the peeler to shave them into fine strips. Add the beetroot shavings to the warm pickling solution. Leave to infuse at room temperature for a minimum of 1 hour. The pickled beetroot will keep in a sterilized sealed container in the fridge for up to 1 month.

While this is happening, make the apple glaze. Stir together the apple juice, apple cider vinegar and sugar in a saucepan. Reduce over a medium heat for about 30–40 minutes or until there's one-sixth left – you should have about 150 ml/5 fl oz/⅔ cup.

To make the burgers, preheat the oven 200°C fan/220°C/425°F/gas mark 7.

Place the sweet potatoes in a roasting pan and roast for 45 minutes. Remove from the oven but leave the oven on. Leave the potatoes until cool enough to touch and then scoop the flesh out of the skins. Place the sweet potato flesh, cooked broad beans, spices, dried chilli flakes, 30 g/1 oz of the flour and some salt to a food processor and blitz until bound together.

Tip the mixture out into a bowl and shape into 4 burger patties with your hands. Dip each patty in the remaining flour to coat the outside all over. Place on a well-oiled baking sheet and roast the burgers for 25 minutes, turning halfway through cooking. Place a slice of goat's cheese on top of each burger and cook for a further 3 minutes.

Assemble the burgers inside your delicious homemade brioche buns, with any condiments you like. I always think a cheeky bit of mayonnaise (especially a herby one) on the bottom bun helps to keep it moist – place the burger on top, then drizzle over some apple glaze and finish with some pickled beetroot and the burger bun lid.

Squash Risotto with Toasted Plum Kernel and Sage Oil

A squash risotto is traditionally paired with amaretti biscuits in Italy for that delicious complementary almond flavour. However, I've always found the partnership a bit sweet for my liking - and I'm not always in the mood for making biscuits with dinner. Here I've used the kernels of stone fruit in an infused oil to go with the risotto instead. I have plums near me, and their kernels also have an almondy flavour, so I use them, but you could use any stone fruit local to you. The raw kernels of stone fruits do contain a mild arsenic, so it's always best to cook them for a few minutes before using to remove this.

SERVES 4

FOR THE KERNEL OIL

10 PLUM STONES (PITS)

100 ML/3⅓ FL OZ/⅓ CUP LOCALLY PRODUCED FLAVOURLESS OIL OF YOUR CHOICE

FOR THE RISOTTO

3 ONIONS

1 LARGE SQUASH (MY FAVOURITE IS UCHIKI KURI)

6 GARLIC CLOVES

120 G/4¼ OZ HARD CHEESE FOR GRATING, PLUS THE RIND

2 CELERY STALKS

8 FRESH SAGE LEAVES AND THEIR STALKS, SEPARATED

LOCALLY PRODUCED COLD-PRESSED OIL OF YOUR CHOICE, FOR ROASTING

300 G/10½ OZ/GENEROUS 1½ CUPS ARBORIO RISOTTO RICE

SEA SALT

To make the kernel oil, cover each plum stone with a tea towel (kitchen cloth) and hit it hard with a rolling pin to break it open. Inside the stones you'll find the kernels which look like pine nuts, set these to one side and discard the smashed stones. Roughly chop the kernels and place them in a small saucepan with the oil. Place over a low-medium heat and wait 2 minutes for the kernels to sizzle and turn a rich brown. Remove from the heat and cool slightly. Transfer to a food processor and blitz the oil and kernels together. Return the oil to the pan off the heat and set aside.

Peel the onions, squash and garlic. Set the vegetables aside and place all the peelings in a medium saucepan with 1.2 litres/quarts of water to make a stock. Chop the rind off the cheese and add to the pan along with one of the celery stalks and one of the onions cut in half. Add the sage stalks and simmer, uncovered, for 40 minutes before straining.

Preheat the oven to 200°C fan/220°C/425°F/gas mark 7.

Meanwhile, halve the squash, discard the insides and save the seeds for another recipe. Chop the squash into cubes, place in a roasting pan, season with salt, drizzle with oil and roast for 20–30 minutes. Remove the squash from the pan and set aside in a bowl.

Finely dice the onion, garlic and remaining celery stalk and add to a large saucepan with a few glugs of oil. Cook over a low-medium heat for 5 minutes, stirring occasionally. Add the rice, stir well and season with salt. Add a ladleful of your vegetable stock, stir and let simmer until it has all been absorbed. Repeat, adding a ladleful of stock at a time, until the rice is just tender; it should take about 20 minutes in total. You might not need to use all the stock (keep leftovers for another dish). Mash half the roasted squash with a fork and then stir this and the remaining squash through the risotto. Season again if needed.

Place the kernel oil over a medium heat, add the sage leaves and sizzle for 1 minute until crisp, then remove from the heat. Serve the risotto with grated hard cheese, sage leaves and drizzled with the kernel and sage oil.

YOUR CHAPTER 2

Now you've made your home sustainable, making your community sustainable is the next step. 50% within 30 miles is one of the main principles of this book, which means aiming to obtain 50% of your produce or goods from within a 30-mile radius of where you live.

This process essentially ensures that we are living our best sustainable lives and investing in our local communities. If we all work on making our own 30 miles amazing, then the whole world will work at its best because nowhere will be forgotten.

Our motto 'connect globally, live locally' (see page 8) is also well illustrated here – by doing our job on a local level and all striving for the same thing, we can ensure that we are all contributing towards a strong and well connected global community.

COMMUNITY

12 DO THE MONKEY DANCE

The human body was made to move. So, please dance, run, swim, do yoga or whatever you need to do to get those pistons fired up. When we are properly energized, we feel more alive and are more likely to get out of the house and be active in our community. Plus, having ease of movement will help us avoid becoming slouch potatoes. The greener revolution is all about getting involved and being proactive – upcycling, interacting with neighbours, cooking in the kitchen, tending to our Charlotte potatoes (my favourite variety) in the community garden.

My body is a temple?

Or more like, your body is a tree. Sorry to bang on about it, but it's a metaphor that works and I'm sticking to it! In order for the whole tree to work, all the little systems have to work too. The leaves have to function for the trunk to fully function. In the same way, for humans, stretching smaller muscles and giving our brains a workout is just as important as strengthening larger muscles. Sometimes it's simply a matter of reconnecting our conscious minds back to the body, which our activities like yoga can help with. Balance and harmony are just as important within our 30 miles as they are in our bodies.

Where's my six pack?

I once tried to lift a piano by myself, which went horribly wrong. I was forced to see a chiropractor that my family had always raved about, although I was dubious and never gave myself fully to her techniques. Eventually, I had a moment of realization and acceptance, and have felt so much better since. One part of her treatment is the aligning of the spine and giving you the skills for better posture. I found that just small adjustments in my everyday actions helped my body become freer.

A little while after this, I saw a photograph of men on a beach from the 1950s. I noticed that their body shapes were all very different to the majority of those we see today – they all had flat stomachs and some had six packs. I realized that they all had great posture, so out

of curiosity I started doing stretches and exercises to improve my own. It was amazing how much better I felt and, not that I was interested in getting a six pack, but my stomach changed as a result of just improving my posture.

I'm sure other factors like a leaner diet and more active lifestyles also made the men from the 1950s fitter. Studies show that keeping active daily by doing things like gardening, taking the stairs and walking can be just as good for your health as spending time in the gym. Spending lots of money on the gym or on plastic surgery to get the perfect body seems like an inefficient use of resources, when actually we could all just improve our posture and be a bit more active in our daily lives instead.

Walking for your ideas

Walking is a seriously underrated sport. Not only is it good for your physical health, but it also clears the mind and lets you interact with the world. I love going for walks, or even a jog, if I want to do a bit of creative thinking. And if you want to really get rooted, go somewhere safe and try walking barefoot through the grass – it's a great way to get back to nature. My grandmother has always said, 'I've never regretted a walk'. But if you want a purpose to your walk, then just do your errands at the same time. It would be great if there was a community list of tasks and you could just log in somewhere online, see which one you fancied doing and incorporate it into your walk.

Plastic-free football

We need to get single-use plastic out of stadiums. This might not be something you've considered before, but can your team be proud to say that they are plastic-free? Ask your team, no matter what sport, to remove single-use plastic via email, social media or however you want. Let the players breathe cleaner air and play better sport.

13 TAKE A BREAK

Whether it's physical, creative or mental, the point of a break is to give you back the energy that you've transferred away. If you spend every moment of every day rushing around, this can often lead to burnout.

The French have it right, as they traditionally take a two-hour lunch break. Every time we eat, our brains have to subconsciously concentrate on digesting food, but if we rush back to work, there's no way we can give this 100%. If you can take a few smaller breaks in your day as well as lunch, then so much the better. The brain can refresh, the body relaxes and we can engage with other humans or nature. Your performance will be better and you'll wake up each day feeling more in control and ready to go.

So please do have a break, but don't have a palm oil-using chocolate brand. Instead, have a sustainable organic tea and a homemade biscuit (see page 79). Your breaktime companion might be water or coffee instead, but for me it's always been tea. I think it's time for a poem:

My Greatest Friend

There is no greater friend than a cup of tea
Nobody's sure if it's a he or a she
But one thing I am sure of is that destiny
Is to always be there for you and me

First thing in the morning and last thing at night
After a long country walk when the sun's shining bright
Sat on a sofa next to a log fuelled fire
Crisp winter nights, the moon rising higher
On summer evenings with sun-kissed light
With a tea in your hand it can only feel right

There's no such thing as a tea that can lie
A tea is honest and nor is it shy
Whether you're laughing or singing or having a cry
Tea leaves and hot water are always on standby

It wakes you up gently
And soothes me greatly
Relax at breakfast with a Darjeeling
Love drunk after a morning shower sing
Put your feet up in the afternoon sipping Assam
In front of the telly eating toast with jam
Or if it's a lovely day and you're feeling particularly gay
Why not have a large pot of fruity Earl Grey

Lapsang souchong smells of pinewood dynamite
Like fireworks under a midnight moonlight
Oolong and yunnan are strange to my Gran
But they relax the body, when stressed for an exam

If you're being healthy and lean
Perhaps the best tea for you is a loose-leaf green
Telling your child a night time lullaby
And slurping on a warm lovely Chai
The garden can offer peppermint and rosehip
Which goes well with a shortbread dip
How about Jasmine or Ceylon
Oh my, the list goes on

Who cares if breakfast comes without juice?
Or there's no Pimm's with tennis and the score is deuce?
I can wait till evening for a glass of wine
Travel the desert without water and still feel fine
I'll get through the day if I forget gin o'clock
Or when my waiter changes my cocktail for a mock
I sometimes need a short shot of coffee
After dinner and taken with a morsel of toffee
And just before bed I love some hot milk and honey
(Drunk every day, I think I would feel funny!)

But a day without tea
Is like a pod missing its pea
(Which I've heard is cold and a little bit empty)
So, wherever you are, or whatever you'll be
There is no greater friend than a cup of tea

SOCIAL MEDIA

If our motto is 'connect globally, live locally', then what's the problem with social media? Well, nothing in theory. Social media has been hugely beneficial for me, both for the business and for my personal life. It truly does connect us to people all around the world and that is an astonishing achievement. We are like the mycelium in the tree (see page 8) but faster, more connected and more complex. But a mycelium's natural job is simple and positive to the survival of the tree. Can we be sure that social media is always having a positive effect?

Just like how there is nothing wrong with a chainsaw if it's being used by the right people with the right training in the right environment cutting down the right tree, there is nothing wrong with social media when used for good in the right way. Now, I'm not saying social media is like a chainsaw, but it has been linked to increased cases of teenage depression and rising levels of anxiety. It can be responsible for sleep patterns becoming out of sync, can serve as a platform for bullying and has spawned concepts like FOMO (fear of missing out), or even JOMO (joy of missing out). The average person spends 2 hours 22 minutes a day on social media, which is scary in itself.

An amazing example of a positive change resulting from the global connection of social media was the #metoo movement. It made the world wake up to the awful reality of sexual harssment towards women. The power that the movement had in bringing a community of people together to feel united and strong enough to speak out was incredible. It ultimately sparked a change in attitudes and made sure individuals were held responsible for their actions.

We have made incredible advances in terms of equality and equal rights for women, especially during the last century, but women are still having to fight for equal opportunity in work and pay. We should take care to remember that just being a follower online sometimes isn't enough to make a real difference in the everyday lives of fellow humankind. To make sure we see change, we need to follow through with actions, implement laws and take steps towards educating people to prevent injustices from happening. So, in any area of life, as well as campaigning online, don't be afraid to stand up for the people around you in real-life situations and practise what you believe in.

Online energy

If everything is energy, then social media is a constant transfer of energy to the online world. Being more aware of how we dole out our precious energy is something we should all work towards. Eating food gives you back the energy that it takes to prepare. Helping your neighbour returns good energy to you in another way. Adding something to the community is rewarding for numerous people in numerous ways. But how much energy that we transfer to our phones or to the virtual world actually comes back to us as positive energy or makes a real difference?

Never forget to give the same energy to the real world as you do to the online world, especially your local community and 30 miles. But equally, remember to actively use social media to connect globally and help stir positive social change toward the next stage. Only by coming at this from all angles can we make sustainability a normalized and successful way of life.

#30green

#30green

#metoo

15 KNOW THY NEIGHBOUR

There are some brilliant free food sharing apps online, like OLIO for example. They work by connecting people who have food they don't want or need with neighbours living nearby who would like it. And it's not just free food that people have bought, you can also find gluts of vegetables that someone has grown or offerings of an abundance of fruit from someone's tree. An even greater achievement is that people are making friends via these apps, and that communities have grown stronger as a result. Our 30 miles is important not just for sustainable food production, but for the friendships and the interactions we have with other people.

Zero waste in everything

Aside from OLIO, there are also apps that allow you to give away or swap anything with your neighbours. If you no longer want or need something, don't throw it away, pass the energy on to someone who wants it.

Communal greenhouses

In many countries across Europe, you can visit beautiful towns with large greenhouses incorporated into their urban landscapes that provide food for the community. If travelling degrades flavour because produce has to be picked too early or isn't stored properly, then surely we should all aim to follow the European example and generate food locally. Urban food landscapes are also great because it means children walk past their food all the time and start to properly understand what it is and where it comes from. Another even better idea would be to modernize the traditional European model and build commercial rooftop hydroponic greenhouses in every city (an incredibly efficient system of growing where mineral-rich solvent is used instead of soil).

Community collaboration

I mentioned back on page 68 (Do The Monkey Dance) that it would be great to have a list of community tasks. I also think it would be brilliant to have a local directory of information for each community that changes with the seasons. I don't just mean a village website, but a modern app or website that is easy to use and says in real time what is happening in your 30 miles. Although the 30-mile concept is aligned with production by the masses rather than mass production, it still involves and promotes technology and advancement. We should celebrate our technological achievements and use them to instil knowledge and skills at a local level.

Not best friends, but neighbours

The greener revolution is about understanding that we are all connected intrinsically, but that doesn't mean we have to be best friends with everyone all the time. We don't even have to have the same political views, practise the same hobbies or like the same films. We can keep our individuality while still acknowledging that we belong to the same species. We don't have to be best friends to know that we are all united in our aim for a sustainable, joyful future.

16 DO LESS BUT BETTER

In my previous book on food sustainability, I talked about the concept of eating less but better-quality meat. Well, we can apply this same concept to sustainability in all areas of life too – just do less but better quality. This is all about changing the habits that we've been taught. So instead of buying more sustainable products, buy less in the first place and upcycle. Start one day by buying 50% less of something than you would normally would (whether it be clothes, food or cleaning products) and make 50% yourself. Then the next week, try two days. Then in another two weeks try three days of buying 50% less. Eventually, you'll be spending 50% less and you'll have some extra money in your pocket. You can apply this to any area in life, but the following instances are perfect examples of this theory in practice in my world:

1) A chef's tomato sauce

I once had a chef come into my pub for an interview. I asked if he could cook Italian food, to which he replied that he had one decent Italian dish (oh dear?!) and he could also make pizzas. Supposedly, he made the best tomato sauce for pizza… he was banging on about sweating red onions, white onions and garlic with olive oil, bay leaves and rosemary, then adding guzzles of red wine and red wine vinegar, oregano and basil. Finally, he mentioned the tomatoes, which were added with a large handful of sugar and salt.

In the pub, tomatoes are 98% of our pizza sauce. We only blend tomatoes, oil, salt and sugar. Nothing else. According to our customers, our pizzas taste great, and I salute the tomatoes for a lot of that success. But for my interviewee, tomatoes were 50% of the sauce. Why should we have to put in more work and ingredients to make an average sauce taste better? Instead, we should grow better quality food so that we can work less, get more nutrition and let the natural flavours speak for themselves.

2) A Christmas menu

At the pub there was a December when we ended up making food for four different menus – the breakfast menu, a Christmas Menu, a roast menu and our normal service. Four menus for one week… it was mad. Looking back, I can see that we tried to do it to offer more choice to more people, but in reality we were only diluting our main product. The next Christmas, I just made our normal menu a bit more 'Christmassy', which meant no pre-bookings, no supplier pre-orders, no deposits, no one forgetting what they pre-ordered three months earlier and all the food was fresh with no waste. Everyone was happier. A perfect example of doing less but better.

3) Children's party food

Does anyone else have childhood memories of kids' parties in village halls where the tables would be groaning under the weight of six different types of sweets, four flavours of crisps, three varieties of biscuits, three plates of white bread ham sandwiches – and the only vegetable was the potato that was used in the spud gun? Looking back now, if we'd had just one massive bowl of homemade popcorn, one amazing bowl of pasta and one homemade cake to share I think we would have all been happy (not to mention healthier). Another example of what's cheaper, sustainable, less and better.

And because we are all part of the same tree, we can't just think about what is enough for ourselves, we have to think: what is enough for the world if 8 billion people are doing the same?

A CIRCULAR ECONOMY

Waste is something that is produced both by humans and in nature. In nature, waste products are usually something else's gain, for example, the way a dung beetle collects manure to roll into balls to feed on or make nests out of, or the way trees absorb the carbon dioxide that we breathe out. But for humans, waste is something that we can choose to use OR let become worthless and even damaging just by the fact of its lingering presence. So, let's rethink the way we see our traditional linear system of plunder > use > discard, and start creating circles like nature instead.

A circular economy is a system that aims to:

☐ Reduce the amount of natural resources extracted in the first place

☐ Ensure that we get the maximum use out of these resources by repairing or recycling them into something else

☐ Ensure that any waste products degrade back to organic matter

There are already some great communities and business collectives out there that have discovered ways to turn their waste into something useful for someone else. For example, clothing offcuts can become insulation, leftover grains from beer making can be fed to pigs, used oil can produce energy in an anaerobic digester. A concern some people have is that a circular economy will take away from

profit or reduce employment by lessening the demand for new resources. But actually, if we slowly shift the balance, it can be a more rewarding system altogether, both for businesses and for families. By creating products or energy out of 'waste' we increase efficiency, increase profit and reduce the demand of public and independent services like waste collection. We are also all encouraged to take responsibility for the circle (not just our own part) and care about what enters it and what leaves it.

We should no longer see ourselves as the 'consumers' but rather the 'transformers'. We don't just consume energy, but we create energy as well, so we should be literally just transforming it into a new kind. Let's change the language in our everyday lives to promote action and remind ourselves that we are not the end user, just another link in the chain.

What can you do?

Repairing, recycling and upcycling possessions is one of the best ways to work towards a circular economy on an individual level (it would be great if we could create a more modern and better servicing industry to make the process of repairing equipment easier and cheaper). You can also help to promote a circular economy by choosing to buy from and do business with companies who use a circular economy model.

In my last book, I used the story of Farmer Spud and Mr Crisp to illustrate how the energy in the classically linear production of crisps can easily be wasted: energy and resources are expended by Farmer Spud to grow the potatoes, which are then bought by Mr Crisp who expends energy transporting them and processing them into crisps with a long shelf life. Your friend picks you up a packet of cheese and onion, which you can't stand, so all that energy and all those resources ultimately end up in the bin. On the next page, I've explained how a circular economy version of crisp production could work.

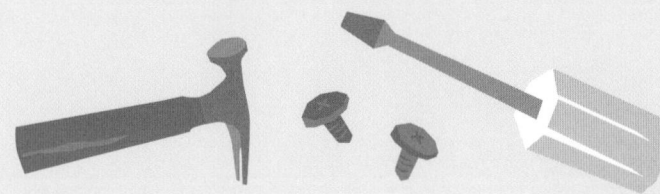

Farmer Spud has grown a wonderful crop of potatoes. Well done Spud.

Mr Crisp comes along and buys two-thirds of the potatoes. The remaining third is sold into the local economy. Farmer Spud is happy.

Mr Crisp's factory is powered only by renewable energy and his solar panels help to heat the water in the factory. The excess steam from crisp production is used to power an energy turbine to make teas for the team.

If they do take the skins off the potatoes, they are given to the local pig farm to feed the pigs instead of being thrown away.

Mr Crisp takes the waste starch from crisp production and it's turned into potato starch packaging, which is fully biodegradable and compostable. The two factories are right next door to each other. The crisps are sent out via vans that run on green electricity or biofuel from waste food production.

20% OFF PICKUP!

Mr Crisp offers 20% off his products if local members of the public come to pick them up from the factory direct, and he offers every ten thousandth customer a free stay at the B&B on Farmer Spud's farm and a guided tour for two hours.

Mr Crisp offsets some of his profits to give Farmer Spud some nitrogen-fixing plants, like broad beans, to help with his soil quality.

The next year, Mr Crisp also buys the beans at a discounted rate and makes them into a special snack under his brand. The rest of the plant is sold to the local community who love to eat root to fruit. Farmer Spud is even happier because he's making more money from his land and he's got really healthy soil.

And the community near Farmer Spud are really grateful to Mr Crisp for offsetting his profit into their community and their 30 miles. They will buy his crisps when they fancy something other than their own homemade ones.

RECYCLING FOR PROFIT

Recycling is not only good for the planet, it can be good for your wallet too. Waste is a by-product but, as the name suggests, it's still a product. The number of discarded by-products around the world must equate to the throwing away of millions, if not billions, of pounds every day. For example, globally there could be a material cost saving of $700 billion just in fast-moving consumer goods like household cleaning products, and $560 billion in the fashion industry.

In the story of Mr Crisp and Farmer Spud (see pages 74–75), the starch from the potatoes was used to make packaging and the steam from cooking the potatoes was turned into electricity. Here the by-products were identified and turned into new products to increase profit and reduce waste. A little time spent thinking creatively, and Mr Crisp now has an exciting circular business model for the future, saving money and reducing waste.

Whether thinking about recycling for profit at home or as part of a business, the first process to go through is the separation of materials. To make life easy for yourself, try to buy goods with minimum packaging that is easily identifiable in the first place. If you do this, you'll have less work to do separating everything out and you won't be confused about what is actually recyclable. The next bit is to find a by-product with a viable market and find a buyer. Have a quick search online, and you'll see there are many apps, websites and social media platforms where you can sell or barter your goods. Some ideas to inspire you are:

☐ Jam jars sold to super trendy bars as glasses

☐ Wine bottles can be sold to make walls (for example in Earthships)

☐ Corks as wedding place holders

☐ Cardboard rolls from toilet paper can be sold as fire lighters

☐ Scrap metal is always worth something

☐ Material or paper can be made into lampshades

☐ Compost is always valuable and very expensive if bought from shops

☐ DVDs and CDs can be cut up and turned into shiny Christmas decorations or hanging shapes that scare pigeons in the garden

☐ Old phones and electrics can be sold online via websites that buy second-hand goods

☐ Old glass can be made into a form of porcelain and then turned into plates and crockery

☐ A company called Precious Plastics have designed a small-scale plastic grinder that makes recycling old plastics into new goods easy for anyone to do. Look them up online: preciousplastic.com

The community as a business

Community can be so much more than just a group of individuals living together in the same area. We have to start seeing the business potential in collaborating with our neighbours. As soon as we realize how much power there is in combining our individual energies, we can make a much more joyful life for ourselves. If recycling was the responsibility of the community instead of the council, it would reduce waste, save money and generate income that could be invested back into your local area however you want. Some ideas below:

☐ As a community, collect materials that are wasted from building projects in the area and then sell them on

☐ Collect all the used cooking oil in your street and sell it to companies who recycle it into energy

☐ Use the money made from your recycling projects to buy a wood chipper, which can then generate compost for all the community and recycle all the scrap wood from packaging or building materials

SLOW FOOD FOR SLOW LIVES

The slow food movement is a global initiative that was first founded in Italy in the 1980s by a man called Carlo Petrini and a group of activists. It celebrates the gastronomic pleasure we gain through taking the time to cook and eat whole and locally sourced foods. Originally a protest against the rise of fast-food culture, it also defends the culture and heritage of regional food traditions. The movement has since grown to champion the benefits of living a slower life in general, encompassing everything from community and the environment, to travel, cities, schools and money. It aims to show how living slow lives lets us go about our day-to-day business in a state of mindfulness and awareness that will lead to a more joyful and connected existence.

Dolce far niente

Mediocrity drives demand, so if we are all rushing around achieving mediocrity, then we feel like we have to do more in order for it to feel good or taste better, when actually the opposite is true. Slow food is amazing because we grow better produce in the first place and therefore have to spend less and do less work to make it tasty. Food is one thing, but how can we all slow life down in general to enjoy it more if we are used to just rushing around?

Stress is one of the leading causes of many diseases including heart disease, obesity, diabetes and depression. Living life in the fast lane puts our bodies and minds in a constant state of 'fight or flight', never allowing us to properly rest. Some people opt to go on expensive retreats, but I think there's a certain irony about having to work so hard to be able to afford an expensive retreat in order to de-stress yourself from working so hard in the first place. I talk later about the art of Transcendental Meditation (see page 107), which is an excellent technique to counter stress, but even just making a simple effort to slow down every day for a little while can help.

The Italians have it summed up perfectly with their phrase, 'dolce far niente!', which translates as 'the sweetness of doing nothing'. The beauty of doing nothing is that it allows us to stop, take stock and just live in the moment. When you take away distractions like worrisome thoughts, reading, social media or television, just you and the moment exist. It's acknowledging that we don't always need something to entertain us or information to enter our minds, but actually just being present in reality and your surroundings is enough. So, don't pay to go on that expensive retreat, just dolce far niente once a day instead.

Four-day work week

Leading a slow life is all about being present, and this includes being present within your community. Slowing your life down and giving time to contribute to the community will allow you to form richer relationships with others, which enhances your sense of general wellbeing. Just an idea, but how about every Friday afternoon, instead of working at their 9–5 up until the last minute, everyone could spend the afternoon pottering about and adding something to society instead – growing food, community cleaning, planting flowers, fixing the roads – bringing the community together while saving the country money. Does anyone really work that hard on Friday afternoon anyway? Only joking...

Cumin-roasted Carrots and Apples with Swede Mash and Elderberries

These are some great autumnal flavours… I think swede mash is my all-time favourite. It has a unique, indulgent yet earthy sweetness that the little pickled elderberry pearls will dance right through. Serve with a helping of beetroot sauerkraut (see page 89) to also help cut through the rich roasted carrots and apples and bring this dish to life.

SERVES 4

12 LARGE CARROTS, TOPS REMOVED AND SET ASIDE

LOCALLY PRODUCED OIL OF YOUR CHOICE, FOR ROASTING

2 TSP CUMIN SEEDS

2 DESSERT APPLES, CORED AND QUARTERED

1 LARGE SWEDE, PEELED AND CUT INTO 2.5-CM/1-INCH CHUNKS

LARGE KNOB (PAT) OF BUTTER (OPTIONAL)

2 BUNCHES OF ELDERBERRIES

SEA SALT

BEETROOT SAUERKRAUT (SEE PAGE 89), TO SERVE

Preheat the oven to 180°C fan/200°C/400°F/gas mark 6.

Halve the carrots lengthways (no need to peel if they're organic but give them a good scrub) and place in a large roasting pan. Drizzle with some oil, sprinkle over the cumin seeds and a little salt. Give them a good mix and then roast for 10 minutes. Add the apples to the roasting pan, mix with the carrots and continue roasting for a further 10 minutes.

Meanwhile, for the mash, put the cubed swede in a saucepan and pour over just enough water to cover. Season with salt and bring to the boil, then reduce the heat slightly and simmer for 10–12 minutes. Drain and mash the swede, adding the butter, or oil if you prefer, and salt to taste.

Pick the elderberries off their stalks and pick out the greenest, softest parts of the carrot tops normally found at the tips to use as a garnish.

Plate up with the mash at the bottom, followed by the roasted carrots and apples, a dollop of beetroot sauerkraut and finally a sprinkling of carrot tops and elderberries to finish.

Hobbly Nobbly Oat Biscuits

Teas

If we make the humble oat the star of the show in a recipe such as this, then its unique character is allowed to shine. You might be able to guess the British variety of commercial biscuit that this recipe is inspired by – their delicious crumbly texture makes them my all-time guilty pleasure. This homemade version is much less guilty and far more delicious.

MAKES 12

150 G/5⅓ OZ/1¼ STICKS ROOM TEMPERATURE UNSALTED BUTTER, PLUS EXTRA FOR GREASING

130 G/4½ OZ/⅔ CUP GOLDEN CASTER (GRANULATED) SUGAR

1 TBSP MILK OF YOUR CHOICE

2 TBSP LOCALLY PRODUCED RAW HONEY

150 G/5⅓ OZ/1 CUP PLUS 2 TBSP SELF-RAISING (SELF-RISING) FLOUR

½ TSP BICARBONATE OF SODA (BAKING SODA)

140 G/5 OZ/SCANT 1½ CUPS ROLLED (OLD FASHIONED) OATS

Put the butter and sugar in a mixing bowl or the bowl of a stand mixer. Cream together using a hand-held electric whisk or the paddle attachment until light and fluffy.

Add the milk and honey and beat again to combine. Sift the flour and bicarbonate of soda into the bowl and fold in. Fold in the oats. Chill the mixture in the fridge for 15 minutes.

Preheat the oven to 150°C fan/170°C/340°F/gas mark 4.

Divide the dough into 25 g/¾ oz pieces and roll into balls. Lightly press each ball into a circle about 0.5-cm/¼-inch deep and space out on 2 greased baking sheets. The sides may crack a little and if this happens just pinch them together. Bake for 15–20 minutes until golden at the edges.

Remove the biscuits from the oven and transfer to a wire rack to cool completely. They will keep in an airtight container for up to 4 weeks.

There is evidence that some companies use microplastics in their teabags, which cannot be broken down and goes directly into our drinks. Therefore, I try to use loose-leaf tea for the best flavour and sustainable points.

A strong black tea with milk is always a winner. The following ideas are more gentle herbal infusions, which are a great way to make use of plants. If picking the leaves yourself, start at the top and take only what you need. For all these flavour suggestions, it's best to make a pot of tea, brew for 5 minutes and use a strainer, at the very least to enjoy the ritual.

RASPBERRY LEAF AND LEMON BALM
6 LEAVES OF EACH PER CUP, RIPPED UP IF TOO BIG

MINT AND LEMON VERBENA
8 LEAVES OF EACH PER CUP

BLACKCURRANT LEAF
6 LEAVES PER CUP, RIPPED UP

FENNEL TOP AND MINT
4 FENNEL TOP SPRIGS AND 6 MINT LEAVES PER CUP

(FENNEL SEED TEA ALSO MAKES A WONDERFUL DIGESTIVE FOR AFTER DINNER, USE 1 TSP PER CUP)

CHAMOMILE FLOWER AND DRIED APPLE
2 FLOWER HEADS AND 3 APPLE SLICES PER CUP

Air dry the chamomile flowers in a dry, dark place for about 10 days or until completely dry before using. To make your own dried apple slices, dehydrate slices of dessert apple at 80°C/175°F in an oven or in a dehydrator for 3–4 hours. These can both be stored in a clean container for up to 3 months.

Spiced Roasted Carrots and Parsnips with Artichoke Purée, Sunflower Seeds and Carrot Top Chimichurri

This dish is exactly what I need on a cold day - sweet parsnips, earthy carrots, creamy artichokes, nutty seeds and fresh, tangy chimichurri, all with a subtle Persian warmth running through. A great dish to celebrate our root vegetables.

SERVES 4

FOR THE CHIMICHURRI

100 G/3½ OZ YOUNG CARROT TOPS

3 TBSP DRIED OREGANO

1 TSP GROUND CUMIN

1 TSP SMOKED PAPRIKA

1 TSP SUMAC

2 GARLIC CLOVES, PEELED

100 ML/3⅓ FL OZ/⅓ CUP MALT VINEGAR

200 ML/6¾ FL OZ/ ¾ CUP LOCALLY-PRODUCED OIL OF YOUR CHOICE, PLUS EXTRA IF NEEDED

PINCH OF SALT

FOR THE ARTICHOKE PURÉE

250 G/9 OZ JERUSALEM ARTICHOKES

50 ML/1⅔ FL OZ/3½ TBSP LOCALLY PRODUCED OIL OF YOUR CHOICE

1 TSP GROUND CUMIN

PINCH OF SALT (OPTIONAL)

FOR THE ROASTED SEEDS

100 G/3½ OZ/¾ CUP SUNFLOWER SEEDS

1 TSP ZA'ATAR

PINCH OF SALT

FOR THE ROASTED CARROTS AND PARSNIPS

12 YOUNG CARROTS, TOPS REMOVED

4 PARSNIPS

2 RED CHILLIES, FINELY CHOPPED

2 TBSP SMOKED PAPRIKA

For the chimichurri, place all the ingredients in a food processor and blitz down into a sauce. You can keep it as chunky as you want, but I like it to be a bit smoother to use as a dressing for this recipe. Add more oil to thin out slightly if needed and set aside. Sauces like this always benefit from being made a little in advance so that the flavours get the chance to meet and mingle.

For the artichoke purée, peel the artichokes tirelessly (reserving the peelings for crisps) or alternatively don't peel them but just give them a really good scrub. Place the artichokes in a large saucepan and cover with water. Bring to a simmer over a medium heat, season with salt and cook for about 45 minutes until the artichokes are tender. Drain but reserve the cooking water. Put the artichokes in a food processor with the oil, cumin, a pinch of salt if needed and a splash of the reserved cooking water and blitz into a beautifully smooth purée. Transfer back to the pan ready to reheat when needed.

Preheat the oven to 180°C fan/200°C/400°F/gas mark 6.

Put the sunflower seeds on a baking sheet. Drizzle with a little oil and sprinkle with the za'atar and a bit of salt. Roast for 10 minutes, then remove and set aside but keep the oven on.

Prep the carrots and parsnips – I normally use organic root vegetables and keep the skins on and give them a really good scrub. Cut them into halves or quarters lengthways depending on the size, trying to keep the final shapes uniform. Place the vegetables in a large roasting pan and sprinkle over the chopped chillies and paprika. Drizzle over some oil and mix thoroughly. Roast in the oven for 15 minutes, then give them a mix and roast for a further 10 minutes.

When you are ready to serve, reheat the artichoke purée and spoon onto serving plates. Pile the root vegetables on top, drizzle with the chimichurri and sprinkle with the roasted sunflower seeds to finish.

81

Homegrown Cannellini Beans with Sourdough

Last year, I tried growing cannellini beans for the first time. I found them to be one of the easiest plants to grow and super rewarding for the fact that they require little growing space for the amount of crop they produce (as is the case with all beans). The trick is to leave them on the plants after a good summer of growth to dry out completely, then store them away for the winter. But as this was my first crop, I was too eager to wait. After taking my prime beans home, I was pondering what to do with these little creamy delights when I remembered a fantastic restaurant in Tuscany that did them so simply - in a rich, herby sauce with a light drizzle of Tuscan olive oil and a chunk of bread… it was stunning. At first, I tried adding other things, but soon realized that the beans shine by themselves. My main course of beans was actually the highlight of my week.

SERVES 4

250 G/9 OZ HOMEGROWN PRIME OR DRIED CANNELLINI BEANS (IF YOU CAN'T GROW THEM YOURSELF THEN ASK A NEIGHBOUR, FRIEND OR FAMILY MEMBER FOR SOME OR FIND A GOOD ORGANIC SUPPLIER. HOMEGROWN BORLOTTI BEANS WOULD ALSO BE DELICIOUS HERE)

1 LARGE ONION

4 GARLIC CLOVES

2 CARROTS, TOPS REMOVED AND KEPT FOR SOMETHING ELSE

1 SPRIG OF ROSEMARY, LEAVES STRIPPED AND STALK RESERVED

10 FRESH SAGE LEAVES

LOCALLY PRODUCED OIL OF YOUR CHOICE, FOR FRYING

2 COURGETTE (ZUCCHINI) STALKS AND LEAVES OR ANY OTHER GREENS YOU HAVE

SEA SALT

GOOD-QUALITY OLIVE OIL, TO SERVE

4 CHUNKS OF SOURDOUGH (SEE PAGES 96-97), TO SERVE

If you are using prime beans, then soak them in plenty of water for at least 2 hours before using. Or, if you are using dried beans, soak them in plenty of water for 8 hours before using. Drain and dry the beans.

Wash and peel the onion and garlic (reserving the cloves and onion) and add the peelings to a large saucepan with the drained beans, 5 cm/2 inches of carrot, the rosemary stalk and 4 of the sage leaves. Cover with double the amount of fresh water. Bring to a simmer and cook, uncovered, for about 40 minutes until the beans are soft. Drain, reserving the cooking water and discarding the aromatics.

Dice the onion, garlic and remaining carrot and finely chop the remaining sage leaves and the rosemary leaves. Add these to a medium saucepan with a few glugs of olive oil and sauté over a medium heat for about 8 minutes, stirring occasionally, until soft. Add the cooked beans and 250 ml/8½ fl oz/1 cup plus 1 tbsp of the cooking water. Season with salt, bring to a simmer and continue to cook for about 5 minutes or until all the water has turned into a creamy sauce.

Five minutes before the end of cooking time, finely slice the courgette stalks and leaves and stir into the beans as they finish cooking. Serve the cannellini beans with chunks of sourdough bread and a little drizzle of the best-quality olive oil.

82

Roasted Stuffed Aubergines with Spiced Spelt and Cider Molasses

I love this recipe - it's so filling and moreish. My take on this Persian classic uses spelt instead of the traditional lamb because I find the flavour works well, but feel free to substitute other grains and adjust cooking times accordingly.

SERVES 4

FOR THE AUBERGINES

2 LARGE AUBERGINES (EGGPLANT)

LOCALLY PRODUCED OIL OF YOUR CHOICE

PINCH OF GROUND CUMIN

SEA SALT

FOR THE STUFFING

LOCALLY PRODUCED OIL OF YOUR CHOICE

2 RED ONIONS, CHOPPED

6 GARLIC CLOVES CHOPPED

BUNCH OF FRESH CORIANDER (CILANTRO), STALKS FINELY CHOPPED AND LEAVES LEFT WHOLE, TO SERVE

1 TSP CUMIN SEEDS

1 TSP GROUND TURMERIC

1 TSP GROUND CORIANDER

1 TSP SMOKED PAPRIKA

½ TSP GROUND CINNAMON

2 RED CHILLIES, CHOPPED

250 G/9 OZ SPELT, PRE-SOAKED FOR 8 HOURS

100 G/3½ OZ LOCALLY GROWN STONED (PITTED) DRIED FRUIT, ROUGHLY CHOPPED

FOR THE DRESSING

CIDER MOLASSES (SEE PAGE 138)

EXTRA VIRGIN OLIVE OIL

TO SERVE

ROASTED PUMPKIN OR SQUASH SEEDS

HOMEMADE YOGURT (SEE PAGE 58)

For the stuffing, pour a few glugs of oil into a saucepan over a medium heat. Add the chopped onions, garlic and coriander stalks and sauté for 3–4 minutes until softened. Add the cumin seeds and fry for 2 minutes before adding all the other spices and the chillies. Fry for a further minute, stirring constantly.

Add the soaked, drained spelt and cook, stirring, for 1 minute. Add 500 ml/17 fl oz/2 cups plus 2 tbsp water and a pinch of salt, stir well and bring to a simmer. Turn the heat down to low and simmer gently, uncovered, for 45 minutes or until the spelt is tender.

Meanwhile, preheat the oven to 180°C fan/200°C/400°F/gas mark 6.

Slice both the aubergines in half lengthways, then score the flesh in a criss-cross pattern, going about 1-cm/⅓-inch deep. Place the aubergine halves in a large roasting pan and drizzle with a few glugs of oil, sprinkle with salt and ground cumin. Rub the cumin and salt into all the scores and roast for 25 minutes.

Remove from the oven (leaving the oven on) and use a fork or spoon to mush the soft aubergine flesh down so there is space to fill them with the stuffing. If there's still a resistance from the aubergine flesh, then cook for a little while longer.

Remove the spelt from the heat when it's cooked, season with more salt if needed and let it stand for a few minutes before mixing in the chopped dried fruit. Pile the spelt stuffing into the aubergine halves and roast for 15 minutes.

For the dressing, combine two-thirds cider molasses with one-third extra virgin or cold-pressed olive oil in a small bowl and whisk well.

Remove the stuffed aubergines from the oven and serve with dollops of yogurt, scatterings of coriander leaves and roasted pumpkin or squash seeds. Drizzle over some cider molasses dressing to finish.

Cauliflower Kedgeree with Smoked Salt and Boiled Eggs

I've been known to eat brinner (breakfast for dinner) many a time… there's something very comforting about it. This dish in particular is so versatile and works as brunch, lunch, dinner or a late-night munch. A kedgeree is traditionally made with smoked haddock, but sustainably sourced smoked haddock may not always be available, so here I've prepared it with a large, white, creamy cauliflower instead. I use smoked salt to get that smoky edge, and a little goes a long way.

SERVES 4

FOR THE CAULIFLOWER

1 LARGE CAULIFLOWER, FLORETS, LEAVES AND STEM SEPARATED

1 TBSP PUMPKIN SEEDS

1 LARGE PINCH OF SMOKED SALT

1 TSP CURRY POWDER

LOCALLY PRODUCED OIL OF YOUR CHOICE, FOR DRIZZLING

FOR THE RICE

2 ONIONS, THINLY SLICED

6 GARLIC CLOVES, DICED

LARGE BUNCH OF FRESH CORIANDER (CILANTRO), STALKS DICED AND LEAVES LEFT WHOLE

1 TBSP CURRY POWDER

1 TSP GROUND TURMERIC

1 TBSP GROUND CUMIN

4 CARDAMOM PODS, SPLIT

1 BAY LEAF

300 G/10½ OZ/1⅔ CUPS BASMATI RICE

SEA SALT

TO SERVE

4 EGGS

BLACK ONION (NIGELLA) SEEDS

HOMEMADE YOGURT (SEE PAGE 58)

Preheat the oven to 180°C fan/200°C/400°F/gas mark 6.

Chop the cauliflower stem into chunks and the leaves into small pieces. Place the cauliflower stem, florets and leaves with the pumpkin seeds in a roasting pan. Sprinkle over the smoked salt and curry powder and drizzle over a little oil. Toss together and then roast for 15 minutes. Set aside until needed.

Add the onions to a large saucepan with a lid along with a drizzle of oil. Cook over a medium heat for about 10 minutes or until the onions are nice and soft.

Add the garlic and coriander stalks. Dice the roasted cauliflower stem and add to the pan along with the spices and bay leaf. Stir and continue cooking for 2–3 minutes. Add the rice, a little sea salt and 600 ml/1¼ pints/2½ cups water and bring to the boil. Stir once to release any rice from the bottom of the pan. Cover the pan with a snugly fitting lid, reduce the heat to low and leave to cook very gently for 12 minutes or until the rice is cooked through and the water has been absorbed.

Meanwhile, boil the eggs for 8 minutes or until hard-boiled. Drain and run under cold water for a minute, then peel and slice in half. Set aside.

Remove the rice pan from the heat, uncover and remove the bay leaf and cardamom pods. Gently fork through the roasted cauliflower florets, leaves and stalks, and place the lid back on top for a further 2 minutes.

To serve, top the rice with the boiled egg halves, scatter over some black onion seeds and dollops of yogurt. Garnish the dish with coriander leaves, if liked.

Chilli and Tomato Jam

You'll be delighted to find this fiery-sweet sauce lurking in your fridge come mid-winter… if it lasts that long. The sourdough soy sauce that's part of this recipe is useful in many other dishes for its umami taste.

MAKES ABOUT 2 KG/4½ LB

3 HOT RED CHILLIES, STALKS REMOVED

8 GARLIC CLOVES, PEELED

2 RED ONIONS, ROUGHLY CHOPPED

2 LARGE THUMB-SIZED PIECES OF FRESH GINGER

1 KG/2 LB 3 OZ RIPE TOMATOES ON THE VINE, SEPARATED

450 G/1 LB/2¼ CUPS GOLDEN CASTER (GRANULATED) SUGAR

150 ML/5 FL OZ/⅔ CUP APPLE CIDER VINEGAR

3 TBSP SOURDOUGH SOY SAUCE (SEE BELOW) OR ANOTHER SUSTAINABLE ALTERNATIVE TO SOY SAUCE

FOR THE SOURDOUGH SOY SAUCE

FEW SLICES OF STALE SOURDOUGH (SEE PAGES 96-97)

For the sourdough soy sauce, soak the bread in just enough water to cover it overnight. The next day, squeeze all the liquid out into a saucepan and discard the bread. Reduce the liquid down by half over a medium heat until it is the consistency of single cream. Set aside until needed.

Put the chillies, garlic, red onions and ginger into a food processor and blitz to a paste. Add the tomatoes and blitz until smooth. You may need to do this in batches.

Transfer the mixture to a saucepan and add the tomato vines, sugar, vinegar and sourdough soy sauce. Bring to the boil, stirring, then reduce the heat and simmer for 40–50 minutes, stirring regularly, until reduced by half.

Leave to cool, then remove the tomato vine and pour into sterilized containers and seal. The jam will keep in the fridge for up to 1 year.

Blackberry, Beetroot and Fennel Seed Chutney

When out looking for blackberries and other hedgerow delights on crisp, clear autumnal days, there is one easy technique to remember - pick the loose berries from the bramble. It doesn't want to be mushy, but if the berry shows no resistance it's a good sign that it is ripe and ready. This chutney works with many foods, whether with duck or venison, salmon or trout, Persian or Scandinavian, Scottish or Moorish… or just in a cheese sandwich.

MAKES ABOUT 1 KG/2 LB 3 OZ

250 G/9 OZ FRESH BLACKBERRIES

250 ML/8½ FL OZ/1 CUP PLUS 1 TBSP APPLE CIDER VINEGAR

500 G/1 LB 2 OZ RAW BEETROOTS, LEAVES REMOVED AND KEPT FOR ANOTHER DISH

4 RED ONIONS, DICED

2 TBSP FENNEL SEEDS

1 STAR ANISE

120 G/4¼ OZ/⅔ CUP MINUS 1 TBSP SOFT LIGHT BROWN SUGAR

Place the blackberries and vinegar in a food processor and blitz down to a smooth purée. Pass the mixture through a sieve (strainer) to remove the blackberry seeds.

Leave the skins on the beetroots but scrub them well. Cut the beetroots into 1-cm/⅓-inch chunks and place in a large saucepan with the blackberry vinegar, red onions, fennel seeds, star anise and sugar. Bring to the boil, then reduce the heat and simmer gently for 30–45 minutes, stirring regularly, until the liquid has almost disappeared.

Leave to cool, then decant into sterilized glass jars, seal and store in the fridge for up to 6 months.

Beetroot Sauerkraut

The colour of this is electric and the beetroot brings a subtle earthy twist to the classic cabbage. It's a versatile and wonderful thing to have in your fridge - great dolloped over grilled fish, served with a curry, a salad or in a sandwich. I also love it with my Cumin-roasted Carrots and Apples with Swede Mash (see page 78).

MAKES ABOUT 600 G/1 LB 5 OZ

300 G/10½ OZ RAW BEETROOTS, PEELED AND FINELY GRATED

300 G/10½ WHITE CABBAGE, ONE LEAF RESERVED AND THE REST THINLY SLICED

1 TBSP CUMIN SEEDS

1 TBSP FENNEL SEEDS

3 TSP SEA SALT

Place the grated beetroot and sliced cabbage in a large bowl. Add the spices and salt, then massage the mixture together, helping to break down the structure of the vegetables. Leave it to stand for 10 minutes.

There should now be some liquid at the bottom of the bowl. Transfer the mixture with the liquid to a sterilized jar that is big enough to leave roughly a 2.5-cm/1-inch gap at the top above the liquid. Push down hard on the vegetables to ensure that the liquid completely covers them. If doesn't, then mix 300 ml/10 fl oz/1¼ cups of warm water with ½ teaspoon of salt and top it up. Cover the mixture with the reserved cabbage leaf. If it needs weighing down, find some clean weights and place them on top of the leaf.

Seal the lid on the jar and leave in a dark place at room temperature for a minimum of 3 days and up to 3 weeks, depending on how fermented you want it (I tend to leave mine for about 1 week). Remember to 'burp' it every 2–3 days by releasing the lid slightly to release the gas. Once ready, keep the sauerkraut in the fridge for up to 2 months.

Fermented Jerusalem Artichokes

A velvety smooth, sweet, nutty, earthy, Jerusalem artichoke purée is a thing of pure beauty. So much so that I'm sure it was Aphrodite who first made it for Ares on the eve of their intense affair. However, there is a slight problem with Jerusalem artichokes that can destroy the most romantic of dinners… the issue of flatulence. Yes, these frisky tubers can give us some gaseous problems. But do not fear! In this recipe yeast is used to slowly break down our beloved vegetable so they are far more digestible. You can then use these fermented artichokes however you like - my choice would be Aphrodite's purée.

SERVES 750 G/1 LB 10½ OZ

750 G/1 LB 10½ OZ RAW ARTICHOKES, SKINS ON

3 GARLIC CLOVES, PEELED

2 TBSP SEA SALT

750 ML/SCANT 1⅔ PINTS/3¼ CUPS LUKEWARM WATER

1 LARGE LEAF, SUCH AS A FIG LEAF, GRAPE LEAF OR CABBAGE LEAF

Cut the artichokes into 2.5cm/1-inch pieces and place in a large jar with the garlic cloves evenly nestled throughout. Mix the salt into the warm water to dissolve, then pour over the artichokes. Cover the artichokes with the large leaf. If it needs weighing down, find some clean weights and place them on top of the leaf. Seal the jar with a lid and leave at room temperature to ferment.

After 2 days, lift the lid slightly to release some gas and 'burp' your artichokes. Continue checking for bubble action and releasing the lid every two days for at least 10 days (and up to 3 weeks).

Once the artichokes are fermented to your liking, transfer the container to the fridge and keep for up to 3 months.

Courgette Stalk Penne with Actual Penne, Courgette, Chilli and Ewe's Cheese

Every now and then I look at a more unusual part of a vegetable or fruit, ponder whether I could eat it and think to myself 'no, I've gone too far'. But for courgette stalks at least, this is certainly not the case. They have a deliciously light courgette flavour, a beautiful texture and they hold their colour very well. You could go all out and serve just courgette stalks as penne (à la courgetti as spaghetti), but I don't want to damage the plant by taking too much, so I bulk this dish out with penne pasta. You can also use squash stalks instead, though be aware that they need a little longer cooking time.

SERVES 4

240 G/8½ OZ DRIED PENNE PASTA

4 LONG COURGETTE (ZUCCHINI) STALKS WITH LEAVES (ABOUT 30 CM/11¾ INCHES), SEPARATED, WITH STALKS SLICED INTO 'PENNE' PASTA AND LEAVES THINLY SLICED

LOCALLY PRODUCED COLD-PRESSED OIL OF YOUR CHOICE

4 GARLIC CLOVES, SLICED

1 RED CHILLI, SLICED

SMALL BUNCH OF FRESH PARSLEY, STALKS AND LEAVES SEPARATED AND BOTH FINELY CHOPPED

2 COURGETTES (ZUCCHINI), GRATED

SEA SALT

FRESHLY-GROUND BLACK PEPPER

GRATED EWE'S CHEESE OR ANOTHER LOCALLY PRODUCED HARD CHEESE, TO SERVE

Bring a large saucepan of salted water to the boil. Add the penne pasta and cook for 4 minutes. Add the penne-shaped courgette stalks to the same pan and cook for a further 5 minutes.

Meanwhile, add a few glugs of oil to a frying pan (skillet) and place over a medium heat. Add the sliced garlic, chilli and parsley stalks. Fry for 2 minutes, stirring, then add the grated courgettes, sliced courgette leaves and a little salt. Stir and cook for a further 3 minutes.

Drain the pasta and courgette stalks, reserving some of the pasta water. Add the penne pasta and stalks to the frying pan along with 5–6 tablespoons of the pasta cooking water. Finally, add the chopped parsley leaves and some more seasoning if needed. Let the sauce bubble away and reduce for about 3–4 W minutes until it has thickened and coated all the pasta.

Serve with plenty of grated ewe's cheese and an extra drizzle of cold-pressed oil.

Truffle-not-Truffle Linguine with Cobnuts and Creamy Cheese

In my book, *30 Ways to a Food Revolution*, I included a recipe for a mushroom ketchup in which the liquid from the mushrooms was extracted, but what to do with the leftover mushrooms was troubling me. Now, I've come up with the perfect solution - we dehydrate the mushrooms and then blitz them up to make a mushroom salt. Try it over steak, poached eggs or in this pasta recipe. It has such an intense taste that it's almost as good as spending £1,000 per kilo on truffle.

SERVES 4

20 PORTOBELLO MUSHROOMS, SLICED

SEA SALT

FOR THE LINGUINE

500 G/1 LB 2 OZ/3¾ CUPS 00 PASTA FLOUR, PLUS EXTRA FOR DUSTING

5 EGGS

1 TBSP OLIVE OIL

SEMOLINA, FOR SPRINKLING

FOR THE SAUCE

100 G/3½ OZ/1 STICK MINUS 1 TBSP BUTTER

FEW GLUGS OF OLIVE OIL

100 G/3½ OZ CHESTNUT (CREMINI) MUSHROOMS, SLICED

50 G/1¾ OZ COBNUTS OR OTHER LOCALLY GROWN NUTS, FINELY CHOPPED

SMALL BUNCH OF FRESH PARSLEY, LEAVES ONLY FINELY CHOPPED

TO SERVE

BUFFALO MOZZARELLA OR LOCALLY PRODUCED CREAMY CHEESE ALTERNATIVE

COLD-PRESSED EXTRA VIRGIN OLIVE OIL

Two days before you want to make this dish, layer the sliced mushrooms in a container just big enough to hold them, sprinkling plenty of salt between each layer. Cover and leave at room temperature for 24 hours. After 24 hours, squeeze all the juice out of the mushrooms which you can use to make mushroom ketchup (see the recipe in my last book!).

Preheat the oven to 90°C fan/110°C/200°F/gas mark ¼ or use a dehydrator to dry out the mushrooms; this should take about 4–5 hours. When ready, blitz them to a coarse powder in a food processor and store in an airtight container (it will keep for up to 6 months).

To make the linguine, add the flour to a mixing bowl and make a well in the middle. Add the eggs and oil to the well, then use your hands to mix the eggs and oil together, gradually mixing with the flour to bring the dough together. Turn the dough out onto a lightly-dusted work surface and knead until smooth. Cut the dough into quarters, place on a plate, cover with a tea towel (kitchen cloth) and rest in the fridge for at least 30 minutes.

Attach a pasta machine to a work surface. Flatten a dough quarter with your hands, dust lightly with semolina and run the pasta through the machine on the thickest setting. Repeat the process with more semolina and taking the thickness down each time you run the pasta through the machine until you reach the second to last thickness (2 mm). If your dough gets a hole, then fold it in half and send through the machine again. Repeat with any remaining dough, then either cut into strips by hand or use the pasta machine attachments to cut the linguine. Leave to hang and dry on a pasta rack or a clean wooden pole for 30 minutes, then use straight away or keep in the fridge until needed.

For the sauce, melt the butter with the oil in a saucepan over a medium heat. Add the mushrooms and nuts and cook for 5 minutes. Meanwhile, cook the linguine in a saucepan of boiling salted water for a few minutes, until al dente. Use tongs to transfer the pasta to the mushroom pan. Add the chopped parsley and a few big sprinkles of mushroom salt. Mix, taste and add more mushroom salt if needed. Serve the linguine scattered with creamy cheese and drizzled with cold pressed extra virgin olive oil.

Ultimate Breakfast Burgers

The real trick with any burger, just like any pizza, is all in the bread - the bun has to be soft, rich and delicious. This brioche-style recipe does the job.

MAKES 6

FOR THE BRIOCHE BUNS

500 G/1 LB 2 OZ/3¾ CUPS PLAIN (ALL-PURPOSE) FLOUR, PLUS EXTRA FOR DUSTING

1½ TSP FINE SEA SALT

50 G/1¾ OZ/¼ CUP GOLDEN CASTER (GRANULATED) SUGAR

130 G/4½ OZ/1¼ STICKS ROOM TEMPERATURE BUTTER, CUBED

25 G/¾ OZ FRESH YEAST OR 12 G/GENEROUS ⅓ OZ DRIED YEAST

130 ML/4½ FL OZ/GENEROUS ½ CUP LUKEWARM WATER

30 ML/1 FL OZ/2 TBSP WHOLE MILK

1 EGG

FOR THE RÖSTIS

2 LARGE POTATOES, PEELED AND GRATED

FEW GRATES OF FRESH NUTMEG

SEA SALT

OIL OF YOUR CHOICE WITH A HIGH SMOKING POINT, FOR OILING AND FRYING

FOR THE BURGERS

6 PORTOBELLO MUSHROOMS

200 G/7 OZ CHEDDAR CHEESE, GRATED

6 EGGS

AQUAFABA AIOLI (SEE PAGE 56)

CHILLI AND TOMATO JAM (SEE PAGE 88)

PICKLES

For the buns, use a stand mixer with a dough hook attachment if you have one. Put the flour, salt, sugar and cubed butter in the bowl of a stand mixer or a mixing bowl and mix with your hands to the consistency of breadcrumbs. In a separate bowl, mix the yeast with the warm water and milk. If using a stand mixer, pour the liquid into the bowl with the flour and knead at a medium speed for 6 minutes. If kneading by hand, mix the yeast liquid into the flour mixture, then turn out onto a lightly floured work surface and knead for 8 minutes. Add the egg, then knead for a futher 2 minutes (in the mixer or by hand). Leave to prove in the bowl, covered with a plate or tea towel (kitchen cloth), at room temperature for about 1 hour or until doubled in size.

Turn the dough out onto a lightly floured work surface and briefly knead it through. Divide the dough into six 120 g/4¼ oz portions and roll into balls, then space out over two lightly oiled baking sheets. Press each dough ball down lightly, then cover each baking sheet with a slightly damp tea towel (kitchen cloth). Prove at room temperature for about 30 minutes or until the rolls have risen to the size you want them to be.

Preheat the oven to 180°C fan/200°C/400°F/gas mark 6. Bake for 10–12 minutes until golden. Transfer to a wire rack and leave to cool.

For the röstis, place the grated potatoes in a sieve over a bowl and squeeze out the liquid. Set the liquid aside for 5 minutes and you should find that a sludgy starch has collected at the bottom of the bowl – this will be your binder. Carefully pour the thinner liquid on top out. Mix the grated potato with 2 tablespoons of the starch, the nutmeg and some salt. Press handfuls of the potato mixture flat between your palms to make 6 thin rösti. Add 0.5-cm/¼-inch of oil to a large frying pan (skillet) over a medium heat. Fry the rösti for 3 minutes on each side until golden, then remove to a plate and set aside.

Increase the oven temperature to 200°C fan/220°C/425°F/gas mark 7.

Place the mushrooms in a roasting pan, drizzle with oil and season with salt. Roast for 10 minutes. Top each mushroom with a handful of grated cheese, add the fried röstis to the pan and roast for another 5 minutes.

Meanwhile, fry the 6 eggs to your liking. To assemble the burgers, halve the buns and spread the bottom with aioli, add a rösti, a cheesy mushroom, some chilli jam, and a fried egg, some pickles and the top bun to finish.

Sourdough

Not all trends are awful. Bucket hats, Crocs, fanny packs, jeggings, skinny ties, denim handbags, sliders, Mini Disc players, segways, deconstructing all types of food… these could all be considered awful trends (or guilty ones). But despite sometimes being given the choice of nine different bread options in a hipster café when all I want is toast, the rise of sourdough bread has only been a good thing.

Sourdough is fantastic for all sorts of reasons. For the bacteria and for humans - nutritionally, flavour-wise and mentally (with the stress relief of slow food). It is the antithesis of processed modern bread because it uses wild yeasts, is slow rising, slow fermented, artisanal and far more delicious. The lactic acids make the gluten more digestible, the sugars slower to absorb and the minerals in the flour more available. These are all something to celebrate.

To avoid having to pay extortionate prices for a loaf in your local deli, use local yeasts and make it yourself. In order to become an artisan, you'll need a 'starter', which you can either make or beg, borrow or steal from a friend or neighbour. A sourdough starter needs attention, and giving it a little time and energy will enhance your bread. In the recipe on the right, I've detailed the easiest and cheapest method to create your own starter. Before long, looking after it will become an easy routine, like looking after a very reliable Tamagotchi… which, by the way, was another dubious trend.

MAKES A STARTER AND 1 LOAF

FOR THE STARTER

ORGANIC WHOLEMEAL (WHOLEWHEAT) FLOUR – I USE THIS AS IT DOESN'T CONTAIN CHEMICALS THAT CAN SOMETIMES AFFECT THE FERMENTATION

LUKEWARM WATER

FOR THE LOAF

500 G/1 LB 2 OZ/GENEROUS 3¾ CUPS WHOLEMEAL (WHOLEWHEAT) FLOUR, PLUS EXTRA FOR DUSTING

15 G/½ OZ/4 TSP SEA SALT

15 G/½ OZ/4 TSP GOLDEN CASTER (GRANULATED) SUGAR

300 G/10½ OZ SOURDOUGH STARTER

280 ML/9½ FL OZ/SCANT 1¼ CUPS LUKEWARM WATER

LOCALLY PRODUCED FLAVOURLESS OIL, FOR OILING THE BOWL

Day 1: To make the starter, place 2 tablespoons of flour in a large, sealable container (about 1 litre/quart capacity). Add 3–4 tablespoons of lukewarm water (only go with the slightly larger amount if the flour you are using will absorb it all). Give it a good mix, seal the container and leave at room temperature somewhere out of the way.

Day 2: Add 1 tablespoon of flour and 2–3 tablespoons of lukewarm water. Give it a stir and reseal. The consistency should be like thick batter.

Day 3: You should begin to see a bit of life, perhaps the odd bubble, perhaps the taint of sourness in the air. If not, don't worry, it just needs another day. Add 1 tablespoon of flour and 2–3 tablespoons of lukewarm water, give it a good stir and reseal.

Day 4: There should definitely be signs of fermentation at this point – probably small holes where the gas has escaped and a vinegary smell. Add 2 tablespoons of flour and 3–4 tablespoons of lukewarm water. Give it a good mix and reseal.

Day 5: By now your fermenting project is well under way and the surface of your starter should be beginning to look more like moon craters. Well, not that big, but it should look alive! Congratulations, your starter is now ready to use. (If it's not looking totally alive, just repeat day 4.) You can now be a bit more generous with the flour, so add 4 tablespoons of flour and 6 tablespoons of lukewarm water. Mix well, seal and leave again. The next day your starter will be ready to use.

Going forward, to feed your starter, remove half and then add ⅓ flour and ⅓ lukewarm water every day with a good whisk (if you are too busy, then you can leave it in the fridge to slow down the fermentation process but you will still need to feed it every couple of days).

To make the loaf, mix the flour, salt and sugar together in a small bowl. In a separate large bowl, mix the sourdough starter and lukewarm water together (if you want it to prove slightly faster, then use body temperature water around 38°C/100°F). Add the flour mixture to the sourdough starter mixture and mix to bring together into a dough.

Turn the dough out onto a lightly floured work surface and knead for a good 8 minutes until smooth and elastic. Lightly oil the same mixing bowl and return the dough to it. Cover with a plate or tea towel (kitchen cloth) and leave to prove in a warm place for at least 3 – 5 hours or until doubled in size.

Preheat the oven to 180°C fan/200°C/400°F/gas mark 6.

Lightly oil a 900 g/2 lb loaf pan. Turn the dough out and knock it back by kneading it through lightly. Shape it into a loaf and place in the oiled pan. Cover with a tea towel (kitchen cloth) and leave the loaf to prove in a warm place again for 1–2 hours, or until it has risen above the top of the pan.

Bake for 35 minutes until risen and golden and it sounds hollow when tapped on the base. Leave the loaf to cool on a wire rack before slicing.

CHAPTER 3

We've covered the local bit of our motto 'connect globally, live locally' by learning how to make our homes and our communities as sustainable and joyful as possible. Now, it's time to focus on what it means to connect globally.

I'm going to talk about ways of living that will help bring us together in our singular aim of creating a greener world, and help us to feel connected to the positive energy of our species, from farming and economics to spirituality, emotion and ideas for the future.

THE WORLD

Once we realize our connection to everything on the planet (and the universe), we can allow ourselves to focus on the changes we need to make.

WHO'S TO BLAME?

The thing is with a revolution, everyday life is hard enough as it is. You've got to make yourself food, do the washing, be a great employee, be a great employer, be sociable, have family time, have enough time for yourself, drive safely, stay fit, make sure you're using condoms... and now I'm telling you that you have to ask yourself a load of questions before you can even start any of this?! Granted, it is tiresome to question whether what we are doing is sustainable at every moment of every day, but we know we have to for the many reasons outlined in this book. So, who on earth do we blame for all this?

In November 2018, my brother was hit by a motorbike and suffered a severe head injury. During the whole process of his recovery, we as a family did not put our energy into blaming anyone. It was clear that the accident was the motorbike driver's fault because he was speeding and on the wrong side of the road, but what good was blame going to do for my brother's recovery? The only thing that really mattered was the strength of our love and putting that toward helping him fight. I genuinely believe that it was this, coupled with my brother's own incredible strength, which allowed him to make the remarkable recovery that he has today.

If everything is connected, and everything is energy, as I said in the beginning of the book, then the concept of blame is just the transference of bad energy to another person. For every step forward, blame makes the whole world take two steps back. Blaming passes on negative energy rather than harnessing positive energy to overcome the problem. Blame can also develop more negative energy within us, creating an unhealthy downward mental spiral.

I don't deny that people should be held responsible for their actions and that they should have to make amends or be put to justice for wrongdoing, but holding someone accountable is different to blaming them. Blame is an energy we throw at someone, which probably won't do any good. But making someone take responsibility or fighting for justice are both positive actions going forward.

Likewise with sustainability, if we can all take personal responsibility for the state of the world by questioning the daily habits in our lives, it will have a hugely positive impact going forwards. If everyone takes responsibility together, then the load is lightened for all. If businesses and producers take responsibility for their ethical actions, then the weight is lifted from the daily consumer. If we can regain our trust in what people are selling, saying and making, then there will come a point when we no longer have to question *everything,* because we can trust that everyone is working towards the same sustainable goal.

So yes, the next stage is going to involve creativity, innovation and a bit of work, because we're going to have to start by asking ourselves questions every day. But when we're on the other side and we've worked out how to live in harmony with the earth, life is going to be so much easier and more lovely...

Who's to blame? Who cares, because my eyes are firmly set on the other side. It's going to be beautiful.

GRIEVY MCNIEVY

It is so important to be able to grieve for what is making us sad. Emotions are truly wonderful things – they're what makes us human – and to express them is absolutely necessary. If they're not allowed to come out, then they can slowly eat away at us, until perhaps one day that big suppressed bubble of emotion might pop... and splatter whoever is nearest... and we don't want that.

If the whole universe is a constant state of transferring energy, then emotions are a hidden energy within us that can be transferred into the world via an expression. To laugh is to express your emotion of happiness, to cry is to express sadness and to speak tender words is to express your love. For people who find this literal expression difficult, emotions can also be expressed through various art forms: joy can be expressed through painting, a cry can be had through a sad poem and love can be most famously expressed through song. We can all be artists with our emotions, so paint, sing, write and enjoy expressing them!

Actually, hang on... don't get happy yet. Back to the importance of grieving.

To be sad is to recognize that something mattered to you, which is great. Because you care and because life means something. I grieve for my brother, even though he has made and is continuing to make an amazing recovery (see page 100) ; I can still grieve because it is a natural emotion that expresses sadness over memories that can be hard to let go of.

I also grieve for the pressure that has been put on people because of climate change. I grieve for the loss of our relationship with nature. For the huge losses of animals and diversity of species. I grieve for the loss of trees, plants and homes destroyed to rear more cattle only to feed our greed. I grieve for the fish that die eating plastic. I grieve for the humans who die because of air pollution or bad diets which they had no choice in. I grieve that the system we're in means that someone can profit from selling unhealthy cheap food in order to afford organic nutritious food for themselves, when others who earn honest livings can only afford to feed their families cheap, non-organic food. I grieve because it is sad. We are allowed to grieve, because it is sad.

But to grieve is also to release the sadness. To release is to let go. To let go is to accept and move on. This too will pass.

22 FORGIVENESS

Forgiveness is everything.

Only by letting go can we make space for joy. Forgiveness can help keep us calm and improve our mental health, because it frees our minds to focus on what is real rather than feeling bitter emotion or resentment. It can make us stronger because we have fully processed and learned from a previous difficult experience. We can then look at the new situation at hand, and intelligently and creatively work out the solution.

To forgive is to unchain ourselves from the people or thing responsible for negativity and to move forward. To forgive means we can reclaim negative energy and use it for something else.

Forgiveness within the context of climate change is not condoning the wrong decisions that have been made. It is not passively letting things slide, and it certainly does not mean forgetting why we are in this mess... but it does mean forgiving those responsible. We have learned from their mistakes and are now ready to change the world to keep the human species alive.

Park and ride

These outer city schemes offer a wonderful solution to reducing the amount of traffic in city centres and keeping the character and atmosphere of a place. More importantly, they also save on our carbon footprint and improve air quality. Perhaps we need more park and cycle schemes as well to take it one step further?

But this concept to me isn't just about reducing traffic. It's also a metaphorical solution for how we can react to problems. Do you remember your parents telling you to count to ten even though your sibling had just done something very annoying? Well, this theory is similar: park your problem and just keep riding. Leave it a while and then see how you feel when you've got some distance and perspective. Come back to it later if you need to, or just keep on riding and leave it behind altogether.

There are so many times in life when you can't see the wood for the trees. You're in the centre of the storm and maybe your vision isn't clear enough to understand what's in front of you. Instead of saying or doing something in the moment that you might regret, allow yourself time and space to work out what's wrong and forgiveness will seem much more viable.

 23 THE NATURAL CYCLES

I think one of the main reasons why life seems to have got harder and harder is because we are working more and more against nature rather than with it.

Philosophers used to think that the real problem with the future would be how us humans would fill our spare time, because robots would be doing most of the work for us. Although this could still happen, something has gone wrong, because even though we are living in an age where we are assisted by more robots and technology than ever before, there is still an extremely heavy focus on working long hours and GDP. Perhaps this is because we're no longer working and using the market to improve the state, instead we're working and using the market to improve our individual lives and accrue individual wealth. A good GDP makes the state look like it has improved health, wealth and happiness all round, but it doesn't take into account the sustainability of our actions. It doesn't ask the question: what have we actually invested into the state for the continued survival of our own species?

The natural world is made up of countless cycles, whether that's in our own bodies, in physics, astrology and biology, the climate and geography, mathematics and economics, or even literature and music. Everything has a natural cycle that we can tap into if only we listen in the first place. Rather than working against these cycles or ignoring them, we should be working with them to create a state that focuses on circular renewal.

So, let's take a break from focusing inwards and look at how we can better work or live with nature in order to improve our lives. We know that we need energy to survive, and the natural world has various sources of energy in abundance. So let's make use of solar, tidal or wind energy rather than plunder sources of non-renewable energy that harm the environment and will eventually run out, leaving future generations in trouble.

Natural cycles in work and play

One of the most obvious natural cycles to occur to roughly half the population is the one that happens monthly within bodies of people who menstruate. Wouldn't it be easier if people were allowed to work or take breaks a little more in rhythm with what is happening in their bodies? Similarly, most people in employment have their own rhythms whereby they are naturally more or less productive at certain times of the day, or times when they just need a rest. Wouldn't it be great if employers could support our natural rhythms by offering us more flexibility to work the hours we choose? This might sound radical but remember, we are the storytellers, so perhaps we just need to change the narrative to allow us to work within our natural cycles.

Of course, working hours are sometimes dictated by the industry you are in. When I began running the pub, I quickly realized that there is a cycle of busy periods interspersed with quiet weeks. Running a pub is hard work, and we give everything to making ours the most sustainable it can be because we believe in the ethos. But to start with, on those quieter weeks it could be quite demoralizing when only one table came in on a night when the team had been there all day, making fridges full of sustainable food from scratch. Perhaps we should have just closed those weeks and said, 'you know what, the energy is flowing a different way'. We could have saved on energy and allowed our team some well-deserved long lie-ins and time off to see friends and family. But the problem with doing this was that the public were not happy to accept that a pub would or could close. Years of preconceptions dictated how we should operate as pub landlords, because our story had already been written. Modern society puts such pressure on everything to be open all the time, and for people to work all the time. If living a slow life and following the slow movement is to become a reality, we need society to change. Remember, we are the storytellers – we can choose to write a new chapter where we go with the flow of nature and work with the cycles, not against them.

I experienced the other side to this when travelling through France one year in mid-August. Frustratingly, it was the time of year when all the locals had decided to take a holiday and close up their shops! It was my fault for forgetting to check how the French locals holidayed. France as a society has accepted this as their natural cycle, knowing that it will be a more fulfilling way of living and better for business in the long run.

Natural cycles in food

Do you ever notice when people say, 'wow, the blossom is looking wonderful this year' or, 'by Jove, what a great year for apples' or 'golly gosh, the blackberries are early this year'? No? Well then perhaps you should listen to *The Archers* on BBC radio 4 more often!

If you have heard these types of expression, then you'll know that every year is different and some years are better for certain crops than others. The natural world, too, regularly needs time off. One year it might be apples, the next it might be roses, the next it might be something else. This autumn was great for plums and blackberries but appalling for elderberries and sloes. So, instead of trying to make sloe gin, I made a wonderful batch of plum vodka instead.

The whole reason I believe we should try to live by the ethos of 50% of produce within 30 miles is because there are some years when blackberries are, to be honest, pretty rubbish. Some years the trendy sloes that every article is telling me to pick are only around for a week. Did you know it takes four weeks for spring to reach the top of Scotland from when it starts in Cornwall? How can we advise people to cook or eat the same thing across the country when this is the case? Especially as the weather is getting less and less predictable. So, what we actually need to do is adapt to the natural rhythms of nature and live moment by moment, according to what is actually happening in the here and now.

Crop rotations

The practice of crop rotation is a great example of how working with a natural cycle brings about the best results. Planting the same crop in the same field year after year depletes the nutrients in the soil and can lead to disease, soil erosion and low crop yield. So, guided by the land, farmers can plant a variety of crops to create rotation cycles and rebalance the soil, allowing everything to grow to the best of its ability.

The problem with this comes when the world demands monoculture or a single variety of crop. If we're not eating the other varieties of crops necessary for healthy soil, then they go to waste and the farmer doesn't get paid for them. In his book *The Third Plate,* Dan Barber theorises that we should align our eating habits with the practice of crop rotations. He created a dish called 'Rotation Risotto', which included a mixture of different crops that a farmer might need to grow in order to give the soil the best nutrient balance. This dish would be different for every farmer on earth, but if we can all aim to eat a more diverse range of foods with the soil in mind, then life will be tastier, healthier and our farmers will be happier.

If this works, then we should each individually think about how we can apply the theory of crop rotation to other aspects of life, like businesses for example. Are there models that we can create whereby a product is made and sold in order to create the next product? Is my waste profitable? What other businesses can use my waste? Are there certain years, months or weeks that I should work harder and some less?

Biodynamic farming

I have to mention this style of farming. When I helped set up a business importing organic wine, I learned a lot about this method and I can tell you the wines it produces are next-level delicious. I wasn't that into wine before – I always thought it was snobby. But when I visited the vineyards, I could relate to it in terms of its sublime taste. Suddenly, the flavour overshadowed the culture.

Biodynamic farming listens to the natural cycles of the earth, so growers will plant, grow, prune and harvest according to when the best moment is for each process. All these practices support natural cycles and the forces of nature. As well as using scientifically supported methods, biodynamic farming also follows more spiritual practices, such as the astrological calendar. I've always thought that if the moon has a huge force on the water on Earth, then why can't it have a huge force on the water in the soils, and therefore affect when fruits or vegetables are most plump and should be harvested. For any sceptics out there, the proof is in the wine.

There are parts of biodynamic farming that may seem a bit weird to people, like the practice of stuffing a cow horn with dung and burying it in the middle of the night, which is meant to attract cosmic forces. This I don't mind, because sometimes in life, the spiritual act means just as much as the science. As a kid, I used to put my wish list for Father Christmas up the chimney – it was a ritual of hope and something I loved doing, whether or not I truly believed those ashes would reach the North Pole. Hope is an energy – a positive energy – and that can only be a good thing when growing. So, if there are a few 'kooky' spiritual acts involved, let us return to our naive innocent minds and make a celebration of them.

Some reviews claim that biodynamic farming produces lower yields than conventional farming, but in the long run it has been shown to achieve better efficiency of production taking into account the amount of energy used. Conventional farming works in a very linear fashion – soil – seed – fertilizer – crop – rotate, and at each stage of the process there is normally a lot of waste, which is disregarded. Biodynamic farming is like a circular economy or sustainable cooking, it uses its own by-products and nature to the best of its ability to create the best outcome for the land. By focusing on the land, not the end product, we can achieve greater yields and greater nutrients for the future. The produce tastes infinitely better, and if we cook sustainably with it, then the balance of energy is far more efficient than using food from conventional farming.

CURING THE WORLD, ONE THOUGHT AT A TIME

I used to work in the bar of a very busy restaurant with over 220 covers. The bar was responsible for making all the drinks, including free coffees for the team of 50. Coffee or tea was a great pick me up and a reason to take a break, but the demand got too much. The management realized they were losing money on their hot drinks and had to limit each member of staff to one per service. This started well, but over the next few months, some people slowly returned to asking for multiple coffees each service again, and everyone else followed suit. But the coffee was a gift from the owners, not part of our wages.

The bar team also had to slice up all the fruits fresh for the drinks that day. One day, a waiter asked for lemon slices with his herbal tea because he had a cold. Sure, I said, it was a nice thing to do and good for morale. The next day, he asked for the lemon slices three times. Sure, he needs to get over the cold I thought. Day three, he asked five times. Day four, six times – and lots of other staff began asking for lemon too. We were having to cut more fruit for staff than for customers. I said no on day five... we were running out of lemons and I could see the management would start rationing these as well! The next day, the guy brought in paracetamol sachets and a grumpy face.

Lessons can be learned from these scenarios; firstly it proves that the actions of many have vastly different consequences to the actions of a few: if everyone in the world thinks to themselves 'I'll just buy one more plastic bottle', that's seven billion more plastic bottles. But if everyone thinks to themselves, I'll grow my own salad today, that's seven billion bits of plastic saved and seven billion more plants producing oxygen. As there is a huge number of our species, when making decisions in life we should always think: 'what happens if everyone does it?'

The second lesson that can be learned is the power of group mentality – how people can be influenced by their peers to act in the same way. I went to a boarding school, and sometimes the older years would come into our rooms and 'lamppost' us, which means to flip the bed over with us in it. It was form of bullying for their own

gratification. As we grew older, our year decided not to carry on this tradition. If we saw other boys do it, then we stopped them. Eventually it stopped altogether. It only takes a few people to make a stand, and eventually the rest will follow suit. Saying no to single-use plastic around friends, at work, in shops will influence others around you.

Positive visualization

Positive visualisation is the act of repeating a thought about a goal you want to achieve in your mind, believing in it and imagining it happen. The mental imagery you generate activates many of the same paths of interconnected nerve cells that link what your body does to the brain impulses that control it. Mental practice makes neural pathways stronger. Positive visualisation that aims to include all the senses, including the feeling after you have achieved what you wanted, is said to increase the effect. Some people like to create physical moodboards of images that they want to achieve, in the belief that looking at it every day will lead to changes. Many well-known people have used positive visualization to achieve success, including athletes, musicians, surgeons and entrepreneurs. Before his success and when he was broke, the actor Jim Carrey would famously visualize becoming a famous actor. He even wrote himself a cheque for 10 million dollars and kept it in his wallet. He eventually went on to sign a movie deal for over 10 million dollars. Obviously, talent, hard work and lots other skills go towards achieving aims and self-fulfillment, but positive visualization is another powerful tool we can use.

Visualizing a sustainable future for yourself, for your loved ones, for your community and for our species can only have a positive effect. It works the other way too, so if a negative thought starts creeping into your mind... get rid of it. Positive thoughts come back to you as positive energy.

Believe in change and it'll start happening.

25 T-ANSCENDENTAL MEDITATION

Learning transcendental meditation, or TM, changed my life. The effect was noticeable pretty much immediately. On odd days when I forget to meditate, I normally look back the next day and think 'oh, that's why I was grumpy' or 'that's why I made that mistake'. But actually, saying it changed my life is quite a simplified way of putting it (it's also a massive cliché). Rather, it might be more accurate to say that TM has made me much more present in my life.

My wife and I had been recommended TM by a variety of new and old friends for a long time, people who we greatly admire and love, so it made sense for us to try it. As trying it involves a four-day course taught by a registered teacher which you have to pay for, it helps to think of it more as an investment in yourself. This might sound a bit corporate and commercial rather than spiritual and enlightening, but I guess everything has to adapt to the systems we're currently in.

The idea is that you get given a mantra that is totally specific to you, and no one else in the world apart from you and the teacher knows your mantra. The mantra has no meaning, so it can't take your thoughts off in a direction. Repeating the mantra in your mind allows you to transcend beyond thought into a state of calm and unbounded awareness. This state of unbounded awareness or pure consciousness is called the 'unified field' in physics, and it's at the source of all mind and all matter, and all matter and all energy emerges from this field – it connects everything.

The techniques you learn on the course make it easy to reach this state in a very natural, effortless way. My wife and I now practise this at home for 20 minutes twice a day. The first question people normally ask is, 'how the heck can you find 20 minutes twice a day?' The simple answer

is you just do. It becomes part of your routine, like brushing your teeth or commuting home from work. And once you get good at it, you can do it anywhere.

Giving your active brain 40 minutes a day to rest is incredibly energy boosting and enhancing. It has allowed me to be calmer, more efficient and more creative, letting me get my work done faster and to a higher quality. I feel truly present in my own thoughts and body, and more aware of what is happening. There have been huge amounts of independent research into the positive effects of TM and these have found that it can:

- ☐ Reduce stress and anxiety
- ☐ Ease depression
- ☐ Increase creativity
- ☐ Increase brain integration
- ☐ Improve relationships
- ☐ Improve general quality of life
- ☐ Improve aspects of physical health, such as blood pressure, reduced rate of heart attack, stroke and death in heart patients

TM is not religious or spiritual, it's an easy technique that anyone can learn. I think forms of meditation like this are an incredibly helpful antidote to the modern phenomenon of uber busy lives and brains. Wouldn't it be even better if we were all taught techniques such as this from a young age? The David Lynch Foundation specializes in getting TM into schools, but I believe it should be on the National Curriculum too.

26 CAPITAL VS CASH FLOW

Back in 1973, economist E. F. Schumacher wrote a book called *Small is Beautiful*. In this he argues that our economy as it stands is unsustainable because natural resources are treated as expendable income, and large industries and large cities only exacerbate the problem. The content is still totally relevant today, which either shows how forward-thinking Mr Schumacher was at the time, or how little we've listened to him. (If we had listened to him 50 years ago, we wouldn't still be using fossil fuels, nor would we be in a climate emergency.)

When you set up or run a business, you have to make sure you get the right balance between capital and cash flow or it can destroy even the best entrepreneur (capital being the amount of liquid assets and cash flow being the income and money spent). Using the right one for the right reasons can make or break a business. For example, working capital can be used to buy or invest in new equipment, expand a business or even help it through rough times. If we rely too heavily on cash flow to do those things, then we eat into profit that isn't even there.

This theory can be applied to our resources on Earth: everything material that humans create comes from a resource on Earth, for example fossil fuels. We currently use these resources to generate cash flow. However, one day our fossil fuels will disappear, and so we need to invest them into renewable energies instead, just as you would invest capital to generate cash flow. Going forward, we should be thinking of fossil fuels as capital not income, investing them into a business that can sustain us to create future cash flow, such as renewable energies or a circular economy.

We have known for a long time, 50 years or more, that we have needed to radicalize our energy systems and build renewable energy sources. So why have we waited until now to panic about it? But all is not lost, we still have reserves of non-renewables that we can use as capital to help build renewable and circular economy systems to generate better cash flow.

Land into money

If we were to look at everything economically, then at the moment it could be said that we are constantly converting land into money – such as destroying rainforests to plant more crops and mass produce more meat, or mining huge plains of tar sands to make cars that run on oil. This money that we're extracting is keeping 'business as usual', in our very linear economic system. We've already spoken about the benefits of a circular economy (see page 73) and how it is a far more sustainable system in the long run. So, if we are to continue using land as a capital asset, we need to find ways of making it circular and reinvesting back into it.

Who is responsible for this? At the top level, I think the United Nations needs to reorganize how we're modelling our global economic structure and what we are doing with our working capital. Perhaps they need to see the world as one nation, one species and one system rather than lots of separate entities with individual aims.

But it's not just all up to the big people, we can each make a difference in our own countries, in our own communities and in our homes. Remember the tree? (See pages 8–9). Its ability to flourish depends on every single part doing its bit, from the huge trunk to the very small leaves and the microscopic mycelium. You are the mycelium and your role is important. Your home is part of the tree and is like a business in itself. Just nurturing your own small patch of land or buying and nurturing a few plants is investing money back into the business.

Order vs freedom

Another natural cycle is the constant swing of the pendulum between order and freedom. As past historic events have shown, for every time of strict order there seems to be a reaction of disorder and happy chaos. What happened after WWII? We had the swinging sixties. Order and chaos are completely different, but we need both of them to keep a state of equilibrium in the world. Like two best friends or partners, one punctual and systematic and the other chaotic and loosey-goosey who balance each other out perfectly.

The economic world we live in is currently extremely ordered, but it is putting too much pressure on centralized industrial systems and large corporations to provide us with everything we need. Let us swing the balance back, continue the cycle and become more free by taking some responsibility for our needs at a local level.

Order gives us the structure and plan we need to create a vision, for which we need capital. After we have created the right systems using our capital, we will have the freedom, or the cash flow, to sustain life.

Green incentive tax reduction

Instead of trying to constantly expand and grow our businesses, how about we encourage them to look inward and become more sustainable instead by creatively using their waste or by-products to generate more income. Wouldn't it be great if even small businesses could get tax incentives to introduce greener practices, like using renewable energies or becoming zero waste? In the long run if all businesses become green it will save them, the country and the world money.

'Any intelligent fool can make things bigger and more complex... It takes a touch of genius – and a lot of courage – to move in the opposite direction.'

E. F. Schumacher

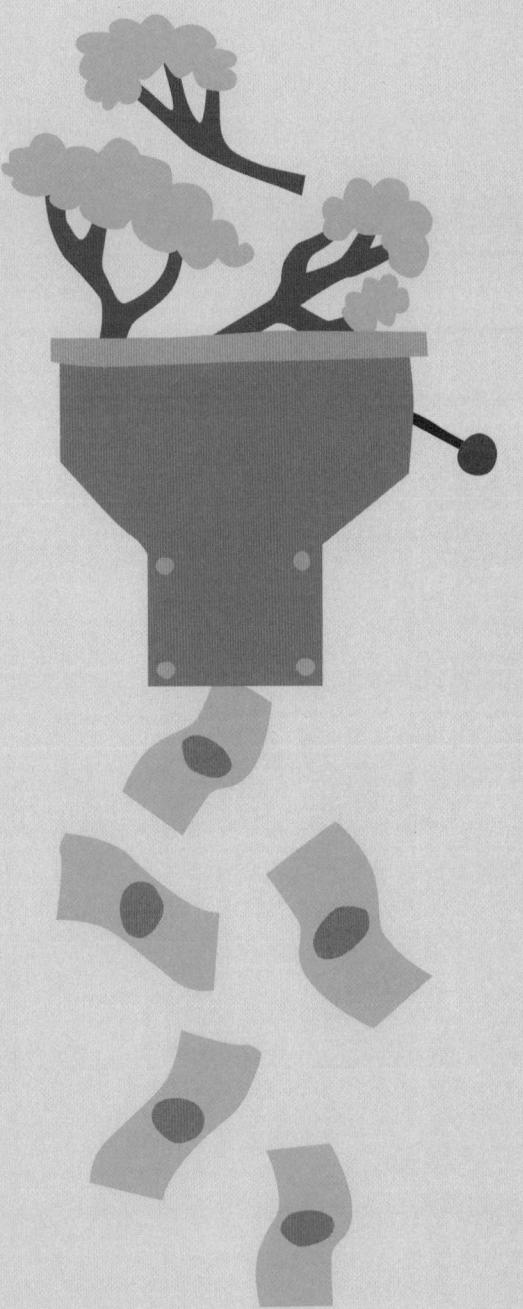

TECHNOLOGY CAN SAVE THE WORLD

I believe in the full unbounded capacity of the human mind and its ability to create. We have created some amazing technology so far – the trick is using it to our advantage and not becoming slaves to it or relying on it for the joyfulness in our lives. Mobiles phones, for example, if used in the wrong way ensure that we are always plugged in to a constant stream of content, which we passively consume rather than nurturing our own creativity and imagination. So let's make sure we use technology in the right way to make the world a better place.

We are living in an exciting new era of collaboration, where the Internet allows us to connect with millions of people around the world to tackle problems, create solutions and learn new information more quickly than ever before. Cutting-edge technology that allows us to be more sustainable is key to the greener revolution – and sustainability boosts the entire economy because it allows produce to be made out of waste or by-products. Let's use our capital to create the best and most efficient technology we can, which in turn can help us generate amazing amounts of cash flow. Here's a list of cutting-edge products and ideas that are focused on providing solutions for a sustainable future:

- ☐ Mushrooms grown from waste coffee grounds
- ☐ Ceramic toilets made from cow manure
- ☐ Reusable coffee cups made from used coffee grounds
- ☐ Eyeware made from cornflour (cornstarch) and microorganisms from used cooking oil
- ☐ Faux leather made from used grape skins after winemaking
- ☐ Textiles and clothes made from orange peel
- ☐ Leather made from pineapple leaf fibres
- ☐ Chipboard made from sugar beet leaves and beet pulp
- ☐ 'Glass porcelain' crockery made from crushed up glass
- ☐ Packaging made from mushrooms
- ☐ Raw milk vending machines
- ☐ Water made into a cleaning product using an electrolyzer (see page 32)
- ☐ Energy made from waste food that has been sent to an anaerobic digester to create electricity and gas
- ☐ An app to make food waste available to your neighbour, such as OLIO, etc.
- ☐ An app that can control the heating at home
- ☐ A house that is made out of 90% waste products like toothbrushes and DVD cases

Technology Can Save the World

28

THERE'S ANOTHER INVESTOR IN THE ROOM

Another metaphor for you. If the Earth is our investor, constantly inputting capital into our business, what would our investor think of the way we are conducting business?

The Chief Scientist of Environmental Health Sciences, Dr. Pete Myers, has warned us of the huge cost of putting chemicals into our food supply. Studies have found that chemical additions to food cost the US health service alone at least $1.2 billion per year. Dr Myers highlights a type of chemical called obesogens that can alter our metabolic processes and predispose some people to gain weight. The scary thing is that obesogens are likely to be found in all our kitchens.

British Professor of Food Policy Tim Lang has said that diet-related diseases are causing global economic losses in the region of $30 trillion (about £23 trillion). That was in 2014. Who's going to tell Earth that one?!

It's awful, but if you were the owner of a business and had just found out statistics like that, you would be pretty nervous about telling your investor. Perhaps if we look at the Earth as this character or even as a person, we can better understand the horrors we are doing to it.

If we have to see everything economically, which the modern capitalist system does, then we are currently converting our land and resources into diseases. This means we need to take more resources (capital) from the Earth to remedy these diseases. So, now that we know the harm chemicals are doing all around the world, let's please our investor and make more circular economy and intelligent business decisions from now on, so that we don't have to be bailed out time and time again. Earth is the largest (and most generous) investor we have in this business. Let's give back what investment and support it has given us.

How to share the load

It's very easy to read big statistics like the ones on the left and feel overwhelmed, so let's make it smaller. Take Tim Lang's statistic of diet-related diseases. If you divide $30 trillion (£23 trillion) by the population on Earth, that's about £3,100 ($4,000) per person. Over an average year then, putting chemicals into our food is costing us about £8.50 ($11) each per day, which doesn't sound quite so bad. If we all individually spend our money on better quality organic food and don't buy into ultra-processed stuff that is pumped full of chemicals, we will be saving the global business huge amounts of money (and we'll be eating more delicious, nutritious food too).

An art installation

Currently, many businesses at all levels of the supply chain operate individually and without any sense of responsibility to the other investor in the room. What this really means is that they can sell you anything and not care where it came from, or how it was made, or who died making it – because they've bought from someone who also has no responsibility over the product. This makes it difficult for us to share the load and buy products that don't harm the Earth. To counter this, I think there should be a monetary value placed on the responsibility knowing where your resources come from. Either tax incentives or penalties for not knowing where your products come from.

I have always wanted to host an art installation inside a supermarket to show where food comes from. When you walk in you would get given some visual simulation goggles. When you look at a product, the goggles would show imagery of where that product actually came from and how it lived its life. If the welfare of your pork was low with awful living conditions, you would see it and think twice about buying it. If you picked up a bag of salad leaves that showed how it had been sprayed with

chemicals and how the plastic it came in was made, you wouldn't buy it. If you picked up a T-shirt and you saw child slave labour, you wouldn't buy it. If everything is more transparent, then this allows consumers to make informed decisions and exercise their free will.

Extinction

The Cretaceous–Paleogene extinction (known as the K Pg extinction) marked the end of the non-avian dinosaurs as well as killing 75% of species on Earth. It was brutal, but it also created huge opportunities for mammals to evolve from the adaptive radiation. Omnivores seemed to survive, unlike their herbivore and carnivore counterparts. The extinction gave birth to new hope and change, which ultimately brought about the dawn of mankind. It's very empowering to think that there has been a total of five mass extinctions on Earth, and that humans exist now because of them. We are the result of natural evolution. The Earth produced us, gave birth to us and raised us.

Dinosaurs lasted for millions and millions of years on Earth, so I think we've got a lot more time here if we play our cards right. Let's not throw it all back in its face, just in case it cuts off our inheritance.

29 WE ARE ANTS ON THE WORLD

Imagine you're lying in a lush green garden on a beautiful summer's day. It's calm and serene and suddenly you feel this movement all the way down near your ankle. You look down and see there is a tiny little ant walking across your leg. That small insect somehow had an effect on you... you changed your action.

Now imagine if there was an army of ants and what chaos or reaction that would create? When you are going about your daily business, remember that we are the ants on this world. Everyone matters. Each person and each family has an effect on the world in their own unique way.

'Man is small, and, therefore, small is beautiful'.

E. F. Schumacher

30   LOVE

The final way to the greener revolution is simply love, because love is the greatest transfer of energy we have.

Love in all its forms is one the best ways to feel connected to this unified field that brings us all together. Sustainability isn't the end goal in itself, sustainability is the end goal because it means love for yourself, for your neighbour and for the world.

Love is not a revolutionary concept, but we can and must express it in many new ways to allow our species to continue.

Love the trees that give us oxygen

Love the crazy jumper that you've patched together with funky material

Love the soil that gives us clothes

Love the cider vinegar that washes your bathroom and makes a salad dressing

Love tea and the well-deserved break

Love the moments in the the day when nothing happens

Love the hashtag that can improve women's rights

Love the water that you save every time you wait for the shower to heat up

Love the people who offer you a shared journey

Love the natural cycles when business is quieter

Love your imagination to turn something unused into something useful

Love the people who are angry and grieving

Love the knowledge that it too will pass

Love human imagination for creating new technology

Love knowing the people who make the things you buy

Love the stories that are about to come

Remember, we are the storytellers and we choose a joyful revolution

Love the small as much as the big

#30green

115

The Whole Venison

If we want to eat meat, then we should be eating a lot less but far better quality. Buying and eating the whole animal is the most sustainable way to do it. Firstly, after the initial upfront cost it will save you money in the long run, whether you share the animal out between friends and family or you have a particularly large freezer. Secondly, you get all the wonderful offal and discarded bits that you don't get in supermarkets, which are super delicious and nutritious. Thirdly, it gives you greater awareness of and control over how many animals you want to kill per year for food, which encourages a more respectful relationship with nature.

For me, if you want to eat meat sustainably, then wild meat is the way to go. Whether it's deer, boar, rabbit, hare, pigeon or partridge; wild meat is delicious and very nutritious. These animals tend to naturally live in balanced environments, promoting diversity and growth. Choosing to consume a variety of wild meat puts a lot less pressure on the meat husbandry industry. Plus, wild animals eat lots of the green stuff, which means their flesh will contain more healthy omegas for you. Ask your local butcher what wild meat options they have – greater demand will create a better supply.

A NOTE ON REWILDING

As we look to the future to restore biodiversity and allow nature to recover, rewilding is certainly on the agenda. It is a process that allows Mother Nature to be back in the driver's seat, to create again a self-sustaining system that puts the ecosystem first. Even in our own backyards can we introduce rewilding principles by allowing grass to grow long in areas, introducing water, and removing any chemicals from gardens.

In her fantastic book *Wilding,* Isabella Tree wrote that "I see rewilding as the webbing that threads through the agricultural landscape...It's a way of recovering the systems upon which farming depends" – and I agree with her whole heartedly.

Venison Neck Ragù with Pappardelle

If you have ever butchered an animal, you'll know how exhausting it is both physically and mentally. If you're asking your butcher for half a venison, or even a whole one, they will really appreciate you asking to keep the neck whole or halved with the bone in. There are some joints of meat that I think are best cooked on the bone, firstly, for the great flavour, but secondly, because it can be a really fiddly job to get all the meat away from the bones - venison neck is one of those.

SERVES 4

FOR THE RAGÙ

½ VENISON NECK, BONE IN (ABOUT 2 KG/4½ LB)

LOCALLY PRODUCED OIL OF YOUR CHOICE WITH A HIGH SMOKING POINT, FOR FRYING

1 ONION, FINELY DICED

1 CARROT, PEELED AND FINELY DICED

1 CELERY STALK, FINELY DICED

6 GARLIC CLOVES, FINELY DICED

8 BROWN MUSHROOMS, FINELY DICED

3 SPRIGS OF FRESH ROSEMARY, LEAVES ONLY, FINELY CHOPPED

250 ML/8½ FL OZ/1 CUP PLUS 1 TBSP RED WINE

1 X 400-G/14 OZ-CAN OF CHOPPED TOMATOES

1 BAY LEAF

2 TBSP WORCESTERSHIRE SAUCE OR SOURDOUGH SOY SAUCE (SEE PAGE 88)

SEA SALT

TO SERVE

360 G/ 11 OZ FRESH PAPPARDELLE PASTA (OR 320 G/10 OZ DRIED)

HARD EWE'S CHEESE, TO SERVE

PURPLE BASIL, TO GARNISH (OPTIONAL)

Preheat the oven to 150°C fan/170°C/340°F/gas mark 4.

Season the venison neck all over with sea salt. Pour some glugs of oil into a large ovenproof saucepan or casserole dish with a lid. Place over a medium–high heat and, when the oil is smoking hot, add the venison neck and briefly sear in the pan until golden on all sides. Reduce the heat down to medium, remove the meat from the pan and set aside. Add the chopped onion, carrot, celery, garlic, mushrooms and rosemary to the pan with another glug of oil. Cook for about 10 minutes, stirring occasionally, until the vegetables are soft and sweet.

Turn the heat up high again, add the red wine and let it bubble away for about 3 minutes until reduced by half. Add the tomatoes, 250 ml/8½ fl oz/1 cup plus 1 tbsp water, the bay leaf and Worcestershire sauce (or sourdough soy sauce). Give it a little stir, then add the seared neck back into the pan. Put the lid on and pop in the oven for 2½–3 hours – the longer the better really – until the meat is so soft that it falls away from the bone.

Remove the pan from the oven and let it cool slightly. Use two forks to remove all the meat from the venison neck bone in the pan – it should be very easy. Place the bone in the bin and mix the meat into the sauce in the pan, breaking it apart further with the forks if needed.

When you're ready to eat, cook the pappardelle pasta as you normally would or following the packet instructions, then drain. Add the pasta into your ragù in the pan and toss together. Place over a high heat again for 3–5 minutes until everything is piping hot.

Add extra seasoning, if needed, and serve with grated hard ewe's cheese and garnished with purple basil, if you like.

Roasted Venison Breast with Mint Sauce

The breast of a venison is one of those parts that normally gets minced or ground because, to be honest, it does take a bit of cooking. But if you can master this simple recipe, it will turn out tender and delicious. This recipe is a take on the Italian peasant food porchetta, which is a rolled and stuffed pork shoulder or belly that has been slowly cooked for hours. It's then either served hot as a banquette centrepiece or sliced and served cold in ciabatta sandwiches.

SERVES 4 AS A MAIN COURSE (OR MAKES LOTS OF SANDWICHES)

2 VENISON BREASTS, BONES REMOVED (ABOUT 1 KG/2 LB 3 OZ)

1 ONION, PEELED AND HALVED

FOR THE STUFFING

1 ONION, FINELY DICED

4 GARLIC CLOVES, FINELY DICED

BUNCH OF FRESH PARSLEY, LEAVES AND STALKS SEPARATED, BOTH FINELY CHOPPED

2 SPRIGS OF FRESH ROSEMARY, LEAVES FINELY CHOPPED

2 TSP FENNEL SEEDS

LOCALLY PRODUCED OIL OF YOUR CHOICE, FOR FRYING

SEA SALT

50 G/1¾ OZ/⅔ CUP HOMEMADE DRIED BREADCRUMBS

50 G/1¾ OZ SOFT LOCALLY GROWN DRIED FRUIT OF YOUR CHOICE, PITTED

FOR THE MINT SAUCE

LARGE BUNCH OF FRESH MINT, DESTALKED AND LEAVES VERY FINELY CHOPPED

1 SHALLOT, VERY FINELY CHOPPED

100 ML/3⅓ FL OZ/⅓ CUP RED WINE VINEGAR OR LOCALLY PRODUCED VINEGAR

30 G/1¾ OZ/¼ CUP GOLDEN CASTER (GRANULATED) SUGAR

To make the stuffing, put the chopped onion, garlic, parsley stalks, rosemary and the fennel seeds in a medium frying pan (skillet) with a few glugs of oil and some salt. Place over a medium heat, stir and then cook for about 6 minutes until the onion is soft. Remove from the heat, add the breadcrumbs and dried fruit and stir together. Stir in a splash of water to bind the mixture together. Set aside.

Lay the venison breasts, skin-side down, on the work surface. Trim off the top and bottom bits of fat and reserve to one side. Season the meat well with sea salt, then spread your stuffing all over the meat. Going from short side to short side, roll the breast up very tightly with the stuffing encased inside. Tie four pieces of butcher's string widthwise around the meat joint at even intervals to hold it together, then tie one long piece of string around the length of it. (If any stuffing has slipped out, you can roast it for 8 minutes in the hot oven alongside the meat.)

Preheat the oven to 200°C fan/220°C/425°F/gas mark 7.

Place the peeled onion halves and venison fat trimmings on the bottom of a large ovenproof saucepan or casserole dish with a lid. Place the rolled venison breasts on top (so they don't directly touch the bottom of the pan) and season the skin with sea salt. Roast for 15 minutes, then remove and add 400 ml/14 floz/1¾ cups of water, pop the lid on the pan, turn the oven down to 150°C fan/170°C/340°F/gas mark 4 and cook for a further 2-3 hours. To check the venison is cooked through, there should be no resistance when a skewer is pushed in. Set aside and leave to rest and cool slightly. If serving as a roast, remove the meat and reduce the cooking juices to make a gravy.

For the mint sauce, put the chopped mint leaves, parsley leaves (from the stuffing ingredients) and shallot in a bowl. Add the vinegar and sugar and mix well to dissolve, adjusting sugar to taste.

Serve the venison cut into slices and drizzled with mint sauce alongside delicious vegetables or refrigerate the whole breast and enjoy it thinly sliced in sandwiches. It will keep in the fridge for up to 5 days.

Kidney and Liver Hogget Meatballs with Brussels Sprout Top Slaw

I have been lucky enough to spend time in the beautiful region of Puglia, south Italy, where they are incredibly localized in the food that they eat. While there, I tried a lot of the offal produced in neighbouring villages. Nothing is wasted, and I had spleen, tripe, heart - the lot. To be honest, I wasn't so keen on all of it. However, I do love liver and kidneys. Nutritionally, they are great for you, and in this recipe they are spiced and mixed with minced venison to give a lovely, delicate introduction to their wonderful flavours.

SERVES 4

FOR THE MEATBALLS

150 G/5⅓ OZ VENISON KIDNEY

150 G/5⅓ OZ VENISON LIVER

300 G/10½ OZ MINCED (GROUND) VENISON

75 G/2⅔ OZ/1 CUP HOMEMADE DRIED BREADCRUMBS

2 GARLIC CLOVES, GRATED

1 TSP GROUND CUMIN

1 TSP GROUND CORIANDER

1 TSP GROUND PAPRIKA

½ TSP GROUND CINNAMON

½ TSP DRIED CHILLI (HOT RED PEPPER) FLAKES

1 EGG

LOCALLY PRODUCED OIL OF YOUR CHOICE, FOR DRIZZLING

FOR THE SLAW

100 G/3½ OZ BRUSSELS SPROUTS TOPS, THINLY SLICED

BUNCH OF FRESH MINT, DESTALKED AND LEAVES FINELY CHOPPED

1 RAW RED BEETROOT, PEELED AND GRATED

2 DOLLOPS OF HOMEMADE YOGURT (SEE PAGE 58)

OLIVE OIL, FOR DRIZZLING

FRESHLY SQUEEZED LEMON JUICE, TO TASTE

SEA SALT

WARM FLATBREADS, TO SERVE

Preheat the oven to 180°C fan/200°C/400°F/gas mark 6.

Very finely chop the kidney and liver or put the meat in a food processor and pulse until finely chopped (makng sure not to purée it). Tip the chopped kidney and liver into a large mixing bowl with the rest of the meatball ingredients. Mix together well to evenly combine, then scoop out portions and form into rounds the size of large golf balls. You should have about 12 large meatballs.

Space the meatballs out in a roasting pan, drizzle with oil and roast for 15–20 minutes until nicely browned and cooked through.

Meanwhile, for the slaw, put the Brussels tops, mint and beetroot in a mixing bowl with the yogurt, a few drizzles of olive oil and lemon juice to taste. Season with salt and then mix together thoroughly.

Serve the hot meatballs with the crunchy slaw and warm flatbreads.

Grilled Sardines with Garlic Scape Salsa Verde and New Potatoes

All garlic plants produce long meandering flowerheads known as 'scapes'. These grow to become pretty tall, and they hold the flower head of the plant. Most growers snap the scapes off to encourage the more sought-after garlic bulb to grow, but these by-products still have a wonderfully mellow garlic flavour with a green, fresh taste. Garlic scapes can be made into all sorts of delicious dishes, but I think they really shine in this garlicky salsa verde paired with the tang of pickled capers, oily sardines and rich, creamy new potatoes.

SERVES 4

8 LARGE FRESH SARDINES, DESCALED AND BUTTERFLIED WITH TAILS LEFT ON

FOR THE GARLIC SCAPE SALSA VERDE

2 GARLIC SCAPES, STRINGY TOPS TRIMMED AND FINELY CHOPPED

BUNCH OF FRESH PARSLEY, STALKS AND LEAVES FINELY CHOPPED

SMALL BUNCH OF FRESH MINT, DESTALKED AND LEAVES FINELY CHOPPED

SMALL HANDFUL OF PICKLED CAPERS OR NASTURTIUM PODS, FINELY CHOPPED

3 TBSP CIDER VINEGAR

1 TBSP DIJON MUSTARD (OR 3 FINELY CHOPPED NASTURTIUM LEAVES)

10 TBSP COLD-PRESSED LOCALLY PRODUCED OIL OF YOUR CHOICE

SEA SALT

FOR THE POTATOES

250 G/9 OZ CHARLOTTE POTATOES OR LOCALLY GROWN NEW POTATOES

BUNCH OF FENNEL TOPS, FINELY CHOPPED

1 RED ONION, THINLY SLICED

100 G/3½ OZ ROCKET (ARUGULA) OR OTHER SALAD LEAVES

LOCALLY PRODUCED OIL OF YOUR CHOICE, FOR DRIZZLING

For the salsa verde, combine the chopped garlic scapes, parsley, mint and pickled capers (or nasturtium pods) in a bowl. Add the vinegar, mustard (or nasturtium leaves) and oil. Season with salt, mix together well and set aside for the flavours to infuse. until needed.

Place the potatoes in a large saucepan filled with cold salted water and bring to a simmer. Cook for about 30 minutes until soft. Drain and leave to cool for 5 minutes, then add the fennel tops, red onion, rocket and a quick drizzle of oil for a dressing. Mix together and set aside.

Preheat the grill (broiler) to high. Place the sardines, skin-side up, onto a grill (broiler) pan and season with salt. Cook under the grill for 3–4 minutes, or until completely cooked through.

Pile the potatoes onto serving plates, lay the grilled sardines on top and drizzle over the garlic scape salsa verde to finish.

Horse Mackerel with Farro and Ground Elder and a Radish Salsa

Ground elder is a wickedly persistent weed, and I find the best way to feel satisfied about getting rid of it is to eat it. The young shoots are the tastiest, somewhere between celery and spinach, and contain lots of goodness. The list of ways to eat our way through this abandoned Roman plant is endless, whether sautéed as greens, simmered in soups or stocks, left whole in salads or even deep-fried as pakoras. So much so that I no longer see it as a weed, but an annual challenge to look forward to. Here I've paired it with nutty whole farro and horse mackerel, which are a delicious alternative to normal mackerel.

SERVES 4

LOCALLY PRODUCED OIL OF YOUR CHOICE, FOR FRYING

1 ONION, FINELY DICED

4 GARLIC CLOVES, FINELY DICED

1 LEEK, FINELY DICED

100 ML/3⅓ FL OZ/⅓ CUP WHITE WINE

250 G/9 OZ/1¼ CUPS WHOLE FARRO, PRE-SOAKED IN WATER FOR 8 HOURS

BUNCH OF FRESH NETTLES, STALKS REMOVED USING GLOVES

BUNCH OF FRESH YOUNG GROUND ELDER, STALKS REMOVED USING GLOVES

4 HORSE MACKEREL, DESCALED, FILLETED AND DEBONED BY YOU OR YOUR FISHMONGER

SEA SALT

FOR THE RADISH SALSA

6 RADISHES WITH THEIR LEAVES, BOTH FINELY CHOPPED

SMALL BUNCH OF FRESH MINT, DESTALKED AND LEAVES FINELY CHOPPED

1 SHALLOT, FINELY CHOPPED

8 ANCHOVY FILLETS, DRAINED AND FINELY CHOPPED

3 TBSP LOCALLY PRODUCED VINEGAR

10 TBSP LOCALLY PRODUCED OIL

BORAGE FLOWERS, TO GARNISH (OPTIONAL)

Preheat the oven to 200°C fan/220°C/425°F/gas mark 7.

Pour some glugs of oil into a large ovenproof saucepan or casserole dish with a lid. Add the diced onion, garlic and leek and place over a low–medium heat. Cook the vegetables for about 5 minutes until just soft, then add the white wine and simmer for about 3 minutes or until reduced by half.

Drain the farro but keep the soaking water. Add the faro to the pan of vegetables with some salt and 500 ml/17 fl oz/2 cups plus 2 tbsp of water. Bring to a simmer, then cover with a lid and transfer to the oven to cook for 30 minutes.

Meanwhile, for the salsa mix together the chopped radishes, radish leaves, mint, shallot and anchovies and mix together with the vinegar and oil. Set aside.

Remove the farro pan from the oven and use gloves to carefully add all the nettles and the ground elder. Season with salt and mix well, then return to the oven for a further 10 minutes with the lid on.

Season each mackerel fillet with sea salt, then remove the pan from the oven once more and place the mackerel fillets one on top of the other (as if they were a whole fish), on top of the farro and cook for a final 10 minutes.

Remove the pan from the oven, spoon the radish salsa over the top and garnish with borage flowers for a lovely cucumber finish if you like. Serve from the pan at the table.

Line-caught Mackerel in Cream Sauce with Broad Beans, Leaves and Stalks

Mackerel is arguably one of the best fish you could ever eat. I have many happy memories of days spent fishing with my family off the Cornish coast, and of the joy of bringing our catch home to gut and barbecue. My memories are of fish with the freshest most delicious flavour and of the feeling of being entirely present with my family, as well as of the pure pleasure of bobbing on the sea.

SERVES 4

4 LINE-CAUGHT MACKEREL, FILLETED AND PIN-BONED

LOCALLY PRODUCED OIL OF YOUR CHOICE WITH A HIGH SMOKING POINT, FOR FRYING

KNOB (PAT) OF BUTTER

2 SHALLOTS, FINELY DICED

80 ML/2¾ FL OZ/⅓ CUP WHITE WINE

200 ML/6¾ FL OZ/GENEROUS ¾ CUP DOUBLE (HEAVY) CREAM

200 ML/6¾ FL OZ/GENEROUS ¾ CUP WATER

200 G/9 OZ FRESH BROAD BEANS (FAVA BEANS)

3 BROAD (FAVA) BEAN STALKS, FINELY DICED

200 G/9 OZ FRESH PEAS

SMALL BUNCH OF FRESH DILL, LEAVES ONLY, CHOPPED

SMALL HANDFUL OF BROAD (FAVA) BEAN LEAVES

SEA SALT

Place the 8 mackerel fillets on a baking tray, skin side up. Oil and salt both sides and turn the grill (broiler) to high.

Heat a glug of oil, the knob of butter and the shallots in a large frying pan (skillet) over a medium heat. Once the shallots are soft add the white wine and reduce by half.

Add the cream and water, season to taste and reduce the heat to low. Once the sauce is at a gentle simmer, add the broad beans, broad bean stalks and fresh peas. Cook for 4-5 minutes until the beans are soft or al dente, then add the dill.

Meanwhile, place the mackerel under the hot grill and cook for 4-5 minutes until the skin is golden.

Spoon the sauce into bowls, place the mackerel fillets on top with any juice left from the baking tray, and garnish with chopped dill and the broad bean leaves.

Here, we have managed to use at least 60% of the broad bean plant, but if we were just using the beans (and some people even like to remove the shells) it would only be about 15%! I also like to keep the broad been casings to add to stock for risottos or thinly slice to make pakoras.

Coronation Fish Chowder with Fig Leaf Milk

Fig leaves have a wonderful flavour and many health benefits. They can be made into teas, stews, stocks, desserts, or just used as a vegetable replacement. In this case, we're going to use them to give a beautiful coconut/cinnamon flavour to the milk in a fish chowder. If you can't find them in supermarkets, then ask friends, family or neighbours who have a fig tree. I've suggested using smoked haddock, but before you buy, make sure the fish is rated either 1 or 2 on the Marine Conservation Society (MCS) guide - this is subject to change and its important to change with it.

SERVES 4

3 FIG LEAVES

600 ML/1¼ PINTS/GENEROUS 2½ CUPS WHOLE MILK

400 G/14 OZ SUSTAINABLY CAUGHT UNDYED SKINLESS SMOKED HADDOCK

100 G/3½ OZ/1 STICK MINUS 1 TBSP BUTTER

LOCALLY PRODUCED OIL OF YOUR CHOICE, FOR FRYING

2 ONIONS, FINELY DICED

SMALL BUNCH OF FRESH CORIANDER (CILANTRO), STALKS AND LEAVES SEPARATED AND BOTH FINELY CHOPPED

2 LEEKS, WHITE AND GREENS SEPARATED AND BOTH SLICED

2 LARGE POTATOES, UNPEELED AND CUT INTO 1.25-CM/½-INCH CUBES

2 TBSP CURRY POWDER

1 TBSP FAVA BEAN (BROAD BEAN) FLOUR OR FLOUR OF YOUR CHOICE

SEA SALT

4 STONED (PITTED) DRIED APRICOTS, OR OTHER LOCALLY PRODUCED DRIED FRUIT, CHOPPED, TO SERVE

100 G/3½ OZ/1¼ CUPS FLAKED (SLIVERED) TOASTED ALMONDS, TO SERVE

Place the fig leaves and milk in a wide saucepan and lay the fish over the fig leaves. Heat the milk to a very gentle simmer and cook for about 2 minutes over a low–medium heat until the haddock is just cooked through. Remove the haddock from the pan with a slotted spoon and set aside on a plate. Keep the pan of milk to one side as well, leaving the fig leaves to infuse the warm milk.

Put the butter and a few glugs of oil into a separate large saucepan over a medium heat. Add the chopped onions, coriander stalks and leek whites and cook for 5 minutes, stirring occasionally, until the onions have softened. Add the cubed potatoes to the pan with the curry powder and continue cooking for a further 5 minutes.

Add the sliced leek greens, then stir in the flour and cook out for 2 minutes. Now slowly begin to add the warm fig leaf-infused milk in batches, stirring well with each addition, until it has all been incorporated into a smooth sauce. Add the cooked fish to the pan, season the sauce with salt and stir to break up the fillets. Cook for a few more minutes until the fish is properly hot.

Serve the fish chowder scattered with chopped dried apricots or other dried fruit, toasted flaked almonds and the coriander leaves to finish.

Sweetcorn Cob Ice Cream

I know it sounds weird, but it works so well. When I gave this ice cream to customers in the pub and asked them to identify the flavour, most thought it was vanilla, crème brûlée or custard. As we look to find new ways to eat more locally and sustainably, perhaps this technique can help reduce the demand for vanilla while still letting us enjoy a similar taste. We can munch the sweetcorn kernels in another dish and all feel better for making use of our cobs.

MAKES ABOUT 2 LITRES/QUARTS

6 SWEETCORN COBS (EARS), KERNELS REMOVED AND SAVED FOR ANOTHER DISH

ROOM TEMPERATURE UNSALTED BUTTER, FOR SMOTHERING

600 ML/1¼ PINTS/2⅔ CUPS DOUBLE (HEAVY) CREAM

600 ML/1¼ PINTS/2½ CUPS WHOLE MILK

8 EGG YOLKS (USE THE WHITES IN ANOTHER RECIPE)

300 G/10½ OZ/1½ CUPS GOLDEN CASTER (GRANULATED) SUGAR

Preheat the oven to 180°C fan/200°C/400°F/gas mark 6.

Place the sweetcorn cobs in a roasting pan, smother with butter and roast for 15–20 minutes.

Add the roasted cobs to a large saucepan with the cream and milk, place over a low–medium heat and simmer gently for about 15 minutes.

Meanwhile, put the egg yolks and sugar in a large bowl and whisk together until smooth. Strain the warm sweetcorn-infused milk into the bowl with the egg yolks and sugar, whisking vigorously with a balloon whisk until combined into a smooth custard. Return the custard to the milk pan and cook slowly over a low heat for about 10 minutes or until thickened. It's ready when it is thick enough to coat the back of a spoon.

Pour into a large, clean bowl and leave to cool completely. Transfer the cooled custard to an ice-cream machine and churn following the manufacturer's instructions until deliciously smooth.

TIP

Corn cobs can be used to flavour stocks and sauces in the same way. For example, just simmer the cobs in the water/broth to infuse a simple leek and sweetcorn soup with flavour. You can also wrap vegetables, fish and meat in cob husks before cooking them.

Parsnip Ice Cream

Raw Strawberry Sorbet

A frosty night can be a gardener's worst nightmare, but a lovely crisp frost can also enhance the natural sweetness of parsnips, which traditionally work so well in savoury wintery dishes. Surprisingly, it's the sweetness of parsnips that makes this ice cream particularly moreish.

Ice-cream churners come in all different shapes and sizes, some great and some awful. The best one I've ever used (that I'm still using today) is my great uncle's, which must date back to the 80s!

MAKES ABOUT 2 KG/4½ LB

3 PARSNIPS, PEELED AND GRATED

600 ML/1¼ PINTS /2⅔ CUPS DOUBLE (HEAVY) CREAM

600 ML/1¼ PINTS/GENEROUS 2½ CUPS WHOLE MILK (OAT MILK MIGHT ALSO BE A NICE ALTERNATIVE HERE)

300 G/10½ OZ/1½ CUPS GOLDEN CASTER (GRANULATED) SUGAR

8 EGG YOLKS (USE THE WHITES IN ANOTHER RECIPE)

Put the grated parsnips in a large saucepan with the cream and milk. Bring to the boil, then turn off the heat and leave to cool and infuse for 30 minutes.

Place the sugar and egg yolks in a large mixing bowl and whisk together with a balloon whisk until slightly fluffy. Pour the parsnip-infused cream mixture through a sieve (strainer) and into the bowl of sugar and eggs. Whisk together vigorously to combine into a smooth custard. Wash the large saucepan and then return the custard mixture back to the clean pan. Cook over a low heat for about 10 minutes until thickened. When ready, the custard should nicely coat the back of a spoon.

Leave to cool completely and then churn in an ice-cream machine following the manufacturer's instructions.

I love ice cream… it's creamy and luscious, but this sorbet recipe is just as decadent in its own refreshing way. It really captures the intense taste of a seriously ripe strawberry - it's literally just like eating them straight off the plant and then some. Find the best and most local strawberries you can for this.

MAKES ABOUT 1 KG/2 LB 3 OZ

700 G/1 LB 9 OZ RIPE STRAWBERRIES, GREEN TOPS REMOVED

200 G/7 OZ/1 CUP GOLDEN CASTER (GRANULATED) SUGAR

FRESHLY SQUEEZED JUICE OF 1 LIME

Place the strawberries in a food processor and blitz down to a purée. Pass the purée through a sieve (strainer), discarding the bits left in the sieve. Set aside.

Put the sugar into a saucepan with 100 ml/ 3½ fl oz/⅓ cup water and place over a low-medium heat. Warm through gently until the sugar has dissolved into the water to make a syrup. Leave to cool.

Mix the sugar syrup and lime juice into the strawberry purée, then churn this mixture in an ice-cream machine following the manufacturer's instructions.

A Glut of Plums

Each plum is different, just like how every person is different. My local Victoria plum tree is bountiful, and its fruits have a great balance between sweet and sour. I also think the kernels are larger in this variety than others. Plum kernels in general have a lovely almondy taste like marzipan, but they do contain a compound that turns into arsenic when consumed raw. Roasting them removes this and also highlights their flavour.

Let's go for a small glut, say 20 plums, which you can make the following three recipes (on this page and overleaf) with.

Plum Sorbet

MAKES ABOUT 1 KG/2 LB 3 OZ

13 PLUMS, HALVED, STONES (PITS) REMOVED AND RESERVED FOR THE RECIPE ON PAGE 132

160–180 G/5⅔–6⅓ OZ/¾–1 CUP GOLDEN CASTER (GRANULATED) SUGAR (DEPENDING ON HOW SWEET YOUR PLUMS ARE)

1 EGG WHITE (USE THE YOLK IN ANOTHER RECIPE) OR 120 ML/4 FL OZ/½ CUP AQUAFABA (CHICKPEA WATER)

Place the plum halves in a medium saucepan with the sugar and 250 ml/8½ fl oz/1 cup plus 1 tbsp water. Bring to a simmer, then turn down the heat so it's gently bubbling, pop a lid on and cook for 10–15 minutes until the plums are soft and oozy.

Place the plums and their cooking water in a food processor and blitz down to a very smooth, runny purée. Sieve (strain) the purée to remove any tiny lumps and then leave to cool completely.

Place the plum purée in a bowl and whisk in the egg white to combine OR add the aquafaba and mix very well. Transfer the mixture to an ice-cream machine and churn following the manufacturer's instructions.

Plum Liqueur

MAKES ABOUT 1 LITRE/QUART

7 PLUMS, HALVED, STONES (PITS) REMOVED AND RESERVED FOR THE RECIPE ON PAGE 132

130 G/4½ OZ/⅔ CUP GOLDEN CASTER (GRANULATED) SUGAR

700 ML/1½ PINTS/SCANT 3 CUPS VODKA

Place all the ingredients in a large, sterilized jar, seal the jar tightly shut and give it a good shake. Leave at room temperature for 2 months, giving the contents a good shake every 3 days.

Strain to remove the plums and store the liqueur in sealed, sterilized bottles at room temperature. It will keep for years and the flavour will only improve over time.

Plum Kernel Crème Brûlée

SERVES 4

20 PLUM STONES (PITS), TAKEN FROM THE PLUMS USED FOR THE RECIPES ON PAGE 130

340 ML/11 ½ FL OZ/1½ CUPS DOUBLE (HEAVY) CREAM

75 G/2⅔ OZ/⅓ CUP PLUS 2 TSP GOLDEN CASTER (GRANULATED) SUGAR, PLUS 4 TSP FOR CARAMELIZING

4 EGG YOLKS (SAVE THE WHITES FOR ANOTHER RECIPE)

Preheat the oven to 180°C fan/200°C/400°F/gas mark 6.

To remove the plum kernels from the stones, cover each stone with a tea towel (kitchen cloth) and hit it hard with a rolling pin to break it open. Inside the stones you'll find the kernels, which look like pine nuts. Discard the smashed stones, place the kernels on a baking sheet and roast for 6 minutes. Leave to cool and then roughly chop.

Place the chopped kernels in a small saucepan with the cream and bring to a simmer for a few minutes. Remove from the heat and use a hand-held stick blender to blitz the kernels and warm cream together (or you can do this in a food processor). Set aside to cool and infuse for 30 minutes.

Reduce the oven temperature to 150°C fan/170°C/340°F/gas mark 4.

Place the pan of infused cream back over a low–medium heat and warm through gently to just hotter than body temperature (38°C/100°F).

Meanwhile, in a large bowl, whisk the sugar and egg yolks together by hand for a couple of minutes until fluffy. Pour the warm cream slowly into the egg mixture while whisking continuously to combine. Sieve (strain) the mixture into a jug (pitcher) and pour into 4 ramekins.

Place the ramekins in a deep roasting pan and then fill the pan with 2.5 cm/1 inch of hot water to make a bain-marie. Cook for 40–45 minutes or until the brûlées are set with a slight wobble in the centre. Leave to cool and chill for a minimum of 2 hours.

When you are ready to serve, sprinkle one level teaspoon of sugar evenly over the surface of each crème brûlée, then caramelize the tops using a chef's blowtorch or by placing under a hot grill (broiler) for a minute. Leave the sugar to set for a few minutes before serving.

Elderflower Champagne

I was lucky enough to start a business importing organic wine, which taught me a lot about what it means to grow organic produce and rely on the land. When you nurture the natural environment where grapes are grown (the terrior), it gives back so much depth of flavour to the produce. So when I see the same branded merlot in every corner store up and down the country, it annoys me that this ubiquitous drink can be churned out so thoughtlessly. Why does each country not create their own tipple, nurture the land and see what unique flavours it brings?

This elderflower champagne is scrummy, light and refreshing. I think it's best to pick the flower heads on a sunny day when they are dry and in full bloom, but also because it makes for a lovely ramble amongst the brambles. It is fermented to about 8% ABV.

MAKES ABOUT 5 LITRES/QUARTS

600 G/1 LB 5 OZ/3 CUPS GOLDEN CASTER (GRANULATED) SUGAR

1 LEMON, CUT INTO QUARTERS, SEEDS REMOVED

8 LARGE ELDERFLOWER HEADS IN FULL BLOSSOM, UNWASHED (THE NATURAL YEASTS WILL FERMENT) BUT ANY BUGS REMOVED

2 TBSP CIDER VINEGAR

SMALL PINCH OF DRIED YEAST (ONLY IF NECESSARY)

Put the sugar and 4 litres/quarts/1 gallon plus 1 cup water in a very large saucepan over a low heat and warm through gently, stirring occasionally, for about 10 minutes until the sugar has dissolved. Leave to cool completely.

Squeeze the juice from the lemon wedges into a large, clean container, add the lemon wedges, elderflower heads, cider vinegar and sugar syrup. Stir well, cover with a muslin (cheesecloth) and leave to steep for 4 days at room temperature.

Check the liquid after 2 days, it should be beginning to look bubbly from the yeast on the flower heads starting to ferment, but if you see no bubbles or foam then add a very small pinch of dried yeast, give it a stir and leave for another 3 days. (It rarely doesn't work, but if no bubbly action happens at all, then you can also just boil the liquid and reduce until it becomes a syrup and use as cordial. Just try the recipe again.)

Strain the mixture through the muslin, bottle it up in sterilized bottles and seal. Leave at room temperature for a further 5–10 days, opening the bottles and then resealing them every other day in order to release a bit of pressure.

When your elderflower champagne is ready it should look fizzy and taste flavourful. Store in the fridge for up to 2 months and drink when in urgent need of refreshment.

Comice Pear Tarte Tatin with Rosehip Chantilly Cream

One of the main aims of sustainability is biodiversity, and to help this along we should all be aiming to utilize as many plant varieties as possible. So, for this tarte tatin, I've chosen to use the variety of pear that grows nearest to me. If everyone does the same thing, we can guarantee that a rich array of fruits and vegetables continue to flourish. So, take this idea forward and use your local variety of sweet apple or pear in your own tatin - you might just find it's up there with the best desserts you've ever eaten.

SERVES 4

FOR THE PASTRY

250 G/9 OZ/1¾ CUPS PLUS 2 TBSP PLAIN (ALL-PURPOSE) FLOUR, PLUS EXTRA FOR DUSTING

2 TBSP GOLDEN CASTER (GRANULATED) SUGAR

250 G/9 OZ/2¼ STICKS COLD UNSALTED BUTTER, CUBED

ABOUT 150 ML/5 FL OZ/⅔ CUP COLD WATER

FOR THE FILLING

110 G/4 OZ/½ CUP PLUS 1 TBSP GOLDEN CASTER (GRANULATED) SUGAR

30 G/1 OZ/¼ STICK ROOM TEMPERATURE UNSALTED BUTTER, CUBED

4 COMICE PEARS, OR YOUR LOCAL VARIETY OF SWEET PEAR, PEELED, CORED AND HALVED

FOR THE CHANTILLY CREAM

200 ML/6¾ FL OZ/GENEROUS ¾ CUP DOUBLE (HEAVY) CREAM

6 TBSP ROSEHIP AND RASPBERRY LEAF SYRUP (SEE PAGE 139)

For the pastry, place the flour, sugar and cubed butter in a large bowl and mix together lightly with your hands – so you can still see chunks of butter. Make a well in the middle and add 100 ml/3⅓ fl oz/⅓ cup of the cold water. Mix to bring together into a dough, adding the remaining water only if needed. Shape the dough into a smooth rectangle, cover with a tea towel (kitchen cloth) and chill in the fridge for 20 minutes.

Roll the dough out on a lightly floured work surface in the direction of the longest sides until it's roughly three times the original length, keeping the edges as straight as you can. Fold the top third into the centre, then fold the bottom third up over that. Give the dough a 90 degree turn, then roll out again to 3 times the length. Repeat these steps 3 more times. Place on a plate, cover with a tea towel (kitchen cloth) and chill in the fridge for at least 30 minutes or until needed.

To make the Chantilly cream, whip the cream and rosehip syrup together to firm smooth peaks, then refrigerate until ready to serve.

Preheat the oven to 200°C fan/220°C/425°F/gas mark 7.

For the tatin filling, add the sugar to a 22-cm/8¾-inch frying pan (skillet) over a medium heat. Stir constantly until the sugar has turned a rich golden colour. Promptly add the butter and continue to stir until melted and combined into a smooth caramel. Turn the heat down to low and place the pear halves, cut insides down, in the pan of caramel. Cook for 5–10 minutes, then turn off the heat. Turn the pears in the pan so the insides are facing up and bunch them close together.

Roll the pastry out on a lightly floured work surface until it's 5 mm/¼ inch thick and cut out a circle 2.5 cm/1 inch bigger than your frying pan. Place the pastry disc on top of the pears in the pan and tuck the pastry in around the edges. Bake for 30 minutes or until the pastry is golden. Remove from the oven, leave to cool for a few minutes, then place a large plate with raised edges over the frying pan and carefully turn the tart out onto the plate. Serve with the rosehip Chantilly cream.

Sunflower Seed and White Chocolate Tart

I came up with this recipe in my search to create a sustainable alternative to traditional frangipane. The method is pretty much the same, but the different ingredients give it its own unique flavour; I think the sunflower seeds bring a peanut-y creaminess and a slight saltiness in a really good way. As with classic frangipane, the end result here is a lot more exciting than the sum of its parts, and so this recipe really deserves a more interesting name than the one I've given it. The name 'frangipane' supposedly comes from the name for the plant that produces the perfume that originally flavoured the dessert. The Latin name for a sunflower is Helianthus, so you could always rename this a Helianthus Tart - as long as you don't know anyone called Helianthus…

SERVES 8

FOR THE PASTRY CASE

250 G/9 OZ/2 CUPS PLAIN (ALL-PURPOSE) FLOUR, PLUS EXTRA FOR DUSTING

75 G/3 OZ/⅓ CUP CASTER (GRANULATED) SUGAR

125 G/4 OZ/1 STICK PLUS 1 TBSP COLD UNSALTED BUTTER, CUBED, PLUS EXTRA FOR GREASING

1 EGG

50 ML/1¾ FL OZ/3½ TBSP COLD WATER

FOR THE FILLING

130 G/4½ OZ/SCANT 1 CUP SUNFLOWER SEEDS

175 G/6 OZ/1½ STICKS ROOM TEMPERATURE UNSALTED BUTTER, CUBED

175 G/6 OZ/¾ CUP PLUS 2 TBSP GOLDEN CASTER (GRANULATED) SUGAR

½ TSP BAKING POWDER

2 EGGS

65 G/2¼ OZ/½ CUP PLAIN (ALL-PURPOSE) FLOUR

100 G/3½ OZ WHITE CHOCOLATE, CHOPPED

3 TBSP LOCALLY PRODUCED JAM (JELLY)

To make the pastry case, sift the flour into a mixing bowl and add the sugar. Add the cold cubed butter and gently rub in with your fingertips until the mixture resembles breadcrumbs. Add the egg and cold water and mix in until the mixture comes together into a dough. Don't overmix or the pastry will become tough. Place the dough on a plate, cover with a tea towel (kitchen cloth) and chill in the fridge for 30 minutes.

Grease a 25-cm/10-inch loose-bottomed tart pan with a little butter. Roll the pastry out on a lightly floured work surface to a thickness of about 0.5 cm/¼ inch and until it is big enough to generously cover your tart pan. Carefully ease the pastry into the pan, making sure you push it into all the corners. Trim away any excess pastry and refrigerate while the oven preheats to 180°C fan/200°C/400°F/gas mark 6.

Prick the pastry base a few times with a fork. Cover the pastry with baking parchment, fill it with baking beans and blind bake for 15 minutes. Remove the paper and beans and bake for a further 5 minutes. Remove the tart case from the oven and leave to cool.

Reduce the oven temperature to 160°C fan/180°C/350°F/gas mark 4.

For the filling, place the sunflower seeds on a baking sheet and roast for 8 minutes until toasted. Remove from the oven but leave the oven on.

Put the butter, sugar and roasted sunflower seeds in a food processor and pulse to roughly chop. Add the baking powder, eggs, flour and white chocolate and blitz again briefly until well combined but still coarse.

Spread the bottom of your baked pastry case with the jam. Pour over the frangipane mixture into the pastry case and spread evenly to the edges.

Bake for 35–40 minutes until the filling has puffed up and is golden on top. Serve the tart warm, ideally with ice cream.

Blackcurrant Leaf Fools with Lavender Shortbread and Roasted Gooseberries

Blackcurrant leaves have a truly wonderful flavour and being organic they're free from pesticides, which can only make this dish tastier (and leaves you free from chemicals too). There's something quite comforting about a fool, perhaps it's the simplicity, perhaps it's the frankness, or maybe it's just the quaint Englishness... either way, use the best organic cream you can find to make this dish really sing and dance on the palate.

SERVES 4 (WITH LEFTOVER SHORTBREADS)

FOR THE SHORTBREAD

110 G/4 OZ/1 STICK ROOM TEMPERATURE UNSALTED BUTTER, CUBED

50 G/1¾ OZ/¼ CUP LAVENDER SUGAR (MADE BY LEAVING LAVENDER SPRIGS IN GOLDEN CASTER/GRANULATED SUGAR IN A SEALED CONTAINER FOR A WEEK), PLUS A BIT EXTRA FOR SPRINKLING

110 G/4 OZ/¾ CUP PLUS 1½ TBSP PLAIN (ALL-PURPOSE FLOUR) PLUS EXTRA FOR DUSTING

FOR THE FOOL

500 ML/17 FL OZ/SCANT 2¼ CUPS DOUBLE (HEAVY) CREAM

80 G/2¾ OZ/½ CUP MINUS 1½ TBSP GOLDEN CASTER (GRANULATED) SUGAR

20 YOUNG BLACKCURRANT LEAVES FROM THE TOP OF THE BUSH

100 ML/3⅓ FL OZ/SCANT ½ CUP HOMEMADE YOGURT (SEE PAGE 58) OR STORE-BOUGHT PLAIN YOGURT

FOR THE ROASTED GOOSEBERRIES

300 G/10½ OZ FRESH GOOSEBERRIES, TOP AND TAILED

2 TBSP GOLDEN CASTER (GRANULATED) SUGAR

FEW SPRIGS OF FRESH MINT, TO SERVE

Start by infusing the cream for the fool. Place the cream, sugar and blackcurrant leaves in a large saucepan and bring to a gentle simmer. Remove from the heat and leave to cool completely.

For the shortbread, place the butter in a mixing bowl, then sift in the lavender sugar (to remove the lavender) and beat in with a wooden spoon until combined. Stir in the flour, then bring the mixture together into a dough with your hands and knead through lightly until smooth.

Roll the dough out on a lightly floured work surface to an 8 x 20 cm/3¼ x 8 inch rectangle, about 0.5-cm/¼-inch thick. Transfer the dough to a greased baking sheet and chill in the fridge for at least 30 minutes.

Preheat the oven to 160°C fan/180°C/350°F/gas mark 4.

Cut the dough into 10–12 rectangles, then space out over the greased baking sheet. Sprinkle the dough with a little more sugar, then bake for 10–12 minutes until the shortbreads are pale golden brown at the edges. Lift them off the baking sheet with a fish slice or spatula and transfer to a wire rack to cool completely.

Turn the oven temperature up to 180°C fan/200°C/400°F/gas mark 6.

Place the gooseberries in a roasting pan, sprinkle with the sugar and roast for 20 minutes until the berries are softened and starting to break down. Transfer the roasted gooseberries to a bowl, mix together lightly and set aside to cool.

To finish making the fools, place the cooled blackcurrant leaf cream and yogurt in a bowl and whisk together until just beginning to thicken but not too stiff. Swirl half the cooled roasted gooseberries into the fool, spoon the mixture into 4 glasses and chill for at least 1 hour or until ready to serve.

Serve the fools with the remaining roasted gooseberries, shortbread biscuits and a few springs of mint. Any leftover shortbreads will keep for up to 1 month in an airtight container.

Molasses

These recipes are very simple, but perfect for preserving fresh produce. Use them in cakes, desserts, drinks, salad dressings, sauces - anything you like. Just think about how many fruits go rotten and are wasted each year... with these recipes an abundance of fruit can be transformed into something useful. If each country or region in the world had its own special molasses variety for visitors to sample, wouldn't it make travelling more wonderful?

Cider Molasses

MAKES ABOUT 500 ML/17 FL OZ/2 CUPS PLUS 2 TBSP

4 LITRES/QUARTS/1 GALLON PLUS 1 CUP LOCAL CIDER

150 G/ ⅓ OZ/¾ CUP GOLDEN CASTER (GRANULATED) SUGAR (OPTIONAL)

Pour the cider into a very large saucepan. Bring to the boil, then reduce the heat slightly and simmer, uncovered, for about 2 hours or until it has reduced by a quarter. If the cider is too dry then you can add sugar to make the molasses thicker.

Leave to cool, then store in sterilized sealed containers at room temperature for up to 6 months.

Blackberry and Elderberry Molasses

MAKES ABOUT 250 ML/8½ FL OZ/1 CUP PLUS 1 TBSP

4 BUNCHES OF ELDERBERRIES

1 KG/2 LB 3 OZ FRESH BLACKBERRIES

120 G/4¼ OZ/⅔ CUP MINUS 3 TSP GOLDEN CASTER (GRANULATED) SUGAR

160 ML/5½ FL OZ/GENEROUS ⅔ CUP CIDER VINEGAR

Pick all the elderberries off their bunches, discarding the stalks, and place them in a food processor with the blackberries. Blitz to a purée, then strain the mixture into a bowl, pushing down with a spoon to extract all the juice.

Add the juice to a saucepan with the sugar, vinegar and 300 ml/10 fl oz/1¼ cups of water. Bring to the boil, then lower the heat slightly and simme, uncovered, for 1 hour or until the molasses is thick and syrupy with the consistency of double (heavy) cream.

Leave to cool, then store in sterilized sealed containers at room temperature for up to 6 months.

Redcurrant Molasses

MAKES ABOUT 250 ML/8½ FL OZ/1 CUP PLUS 1 TBSP

1 KG/2 LB 3 OZ REDCURRANTS

120 G/4¼ OZ/⅔ CUP MINUS 3 TSP GOLDEN CASTER (GRANULATED) SUGAR

60 ML/2 FL OZ/¼ CUP CIDER VINEGAR

As redcurrants are tarter, less acidity is needed for this recipe. Place the redcurrants in a food processor and blitz to a purée. Strain the purée into a bowl, pressing down with a spoon to extract all the juice.

Add the juice to a saucepan with the sugar, vinegar and 300 ml/10 fl oz/1¼ cups water. Bring to the boil, then reduce the heat slightly and simmer, uncovered, for 1 hour or until the molasses is thick and syrupy with the consistency of double (heavy) cream.

Leave to cool, then store in sterilized sealed containers at room temperature for up to 6 months.

Rosehip and Raspberry Leaf Syrup

Never mind the flowers, for me the real jewels of the rose bush are the rosehips. They have a delicious flavour and contain ridiculous amounts of vitamin C that seeps into water when they are simmered. Adding a little sugar helps us to preserve their goodness for the whole winter. The raspberry leaves give an additional flavour twist, along with another vitamin and mineral boost.

MAKES ABOUT 2 LITRES/QUARTS

1 KG/2 LB 3 OZ ROSEHIPS, LEAVES AND STALKS REMOVED

30 RASPBERRY BUSH LEAVES

600 G/1 LB 5 OZ/3 CUPS GOLDEN CASTER (GRANULATED) SUGAR FOR EVERY 1 LITRE/QUART OF LIQUID

Fill a large saucepan with 1.2 litres/quarts of water and bring to a simmer. Put the rosehips in a food processor and pulse a few times to finely chop them into very small pieces, taking care not to blend and turn them into a mush. Immediately transfer the chopped rosehips to the pan of simmering water. Bring to the boil, then add the raspberry leaves. Reduce the heat slightly and simmer uncovered for 20 minutes.

Strain the liquid through a double layer of muslin (cheesecloth) and leave the muslin over the pan for 30–45 minutes to allow the liquid to fully drain. You can wash the muslin and strain again if you want.

Measure the liquid and return it to the pan. For every 1 litre/quart of liquid, add 600 g/1 lb 5 oz/3 cups of sugar. Simmer over a medium heat for about 10 minutes, stirring now and again, until all the sugar has dissolved.

Leave to cool and then pour the syrup into sterilized glass bottles and seal. The syrup will keep at room temperature for 6–12 months.

A

apples: celeriac top and
  apple soup   48
cumin roasted carrots
  and apples   78
aquafaba aioli   56
aubergines (eggplants):
  baba ganoush   54
  crispy aubergine and honey
    bruschetta   40
  lentil Bolognese   59
  roasted stuffed aubergines
    with spiced spelt   84

B

baba ganoush   54
beetroot: beetroot sauerkraut   88
  blackberry, beetroot and fennel seed
    chutney   88
  Brussels sprout top slaw   120
  spiced beetroot yogurt   50
  spiced broad bean burgers   62
biscuits, hobbly
  nobbly oat   79
blackberries: blackberry and
  elderberry molasses   138
  blackberry, beetroot and
    fennel seed chutney 88
blackcurrant leaf fools   140
Bolognese, lentil   59
bread: cannellini beans
  with sourdough   82
  coffee ground bread   45
  crispy aubergine and honey
    bruschetta   40
  sourdough   96–7
  spelt, seaweed and
    rosemary bread   43
  tomato leaf focaccia   42
brioche: ultimate breakfast burgers   95
broad (fava) beans: line-caught
  mackerel in cream sauce with broad
  beans   124
  spiced broad
    bean burgers   62
broths *see* soups & broths
bruschetta, crispy aubergine
  and honey   40

Brussels sprout tops: Brussels sprout
  top slaw   120
  Brussels sprout tops and sea greens
    with fermented broad
    bean broth   52
burgers: spiced broad
  bean burgers   62
  ultimate breakfast burgers   95

C

cabbage: beetroot sauerkraut   88
camelina seed crackers   44
cannellini beans with sourdough   82
carrot top chimichurri   81
carrots: cumin roasted carrots
  and apples   78
  spiced roasted carrots and parsnips 81
cauliflower kedgeree   86
celeriac top and apple soup   48
champagne, elderflower   133
Chantilly cream, rosehip   134
cheese: barbecued leeks with Cheddar,
  poached eggs and tarragon   55
  courgette stalk penne with
    actual penne, courgette, chilli
    and ewe's cheese   93
  spiced broad bean burgers with
    goat's cheese   62
  truffle not truffle linguine   94
  ultimate breakfast burgers   95
chickpea Scotch eggs   56
chillies: chilli and tomato jam   88
  courgette stalk penne   93
chimichurri, carrot top   81
chocolate: sunflower seed and white
  chocolate tart   136
chowder, Coronation fish   126
chutney, blackberry, beetroot
  and fennel seed   88
cider molasses   138
  roasted stuffed aubergines with   84
cobnuts, truffle not truffle
  linguine with   94
coffee ground bread   45
Coronation fish chowder
  with fig leaf milk   126
courgettes: courgette stalk penne
  with actual penne, courgette, chilli
  and ewe's cheese   93
  dandelion and courgette pakoras 50

crackers, camelina seed   44
cream: blackcurrant
  leaf fools   140
  line-caught mackerel
    in cream sauce   124
  plum kernel
    crème brûlée   132
  rosehip Chantilly cream   134
croquettes, dulse and oat   48

D

dandelion and courgette pakoras   50
dulse and oat croquettes   48

E

eggs: barbecued leeks with Cheddar,
  poached eggs and tarragon   55
  cauliflower kedgeree with
    boiled eggs   86
  chickpea Scotch eggs   56
  ultimate breakfast burgers   95
elderberries: blackberry and
  elderberry molasses   138
  cumin roasted carrots and
    apples with swede mash and
    elderberries   78
elderflower champagne   133

F

farro, horse mackerel with   123
fermented broad bean broth   52
fermented Jerusalem artichokes   89
fig leaf milk   126
fish: Coronation fish chowder   126
  grilled sardines with garlic
    scape salsa verde and
    new potatoes   122
  horse mackerel with farro
    and ground elder and
    a radish salsa   123
  line-caught mackerel in cream
    sauce with broad beans, leaves
    and stalks   124
focaccia, tomato leaf   42
fools, blackcurrant leaf   140